Carbon Nation

CultureAmerica

Erika Doss
Philip J. Deloria
Series Editors

Karal Ann Marling
Editor Emerita

Carbon Nation

Fossil Fuels in the Making of American Culture

Bob Johnson

 University Press of Kansas

© 2014 by the University Press of Kansas

All rights reserved

A version of Chapter 3 originally appeared under the title "An Upthrust into Barbarism: Coal, Trauma, and the Origins of the Modern Self" in the *Journal of American Culture*, © 2010 Wiley Periodicals

Published by the University Press of Kansas (Lawrence, Kansas 66045), which was organized by the Kansas Board of Regents and is operated and funded by Emporia State University, Fort Hays State University, Kansas State University, Pittsburg State University, the University of Kansas, and Wichita State University

Library of Congress Cataloging-in-Publication Data

Johnson, Bob, 1967–
    Carbon nation : fossil fuels in the making of American culture / Bob Johnson.
        pages cm
    Includes bibliographical references and index.
    ISBN 978-0-7006-2004-3 (hardback)
    ISBN 978-0-7006-2520-8 (paperback)
    1. Fossil fuels — Social aspects — United States—History. 2. Energy consumption —Social aspects—United States—History. 3. Energy industries—United States—History. 4. United States — Economic conditions. 5. United States—Environmental conditions. 6. United States — Civilization. I. Title.
    HD9502.U52J653 2014
    306.30973 — dc23

                                                        2014026653

British Library Cataloguing-in-Publication Data is available.

Printed in the United States of America

10 9 8 7 6 5 4 3 2 1

For my mom, my dad, Theresa, and Ava

Contents

But don't you have to have the captain's — or someone's — permission to visit the stokehole?

—*Eugene O'Neill,* The Hairy Ape

Acknowledgments

To write this book, I took on deep debts, most of which will never be adequately repaid. During the last several years, I leaned on my family, friends, peers, and advisors for things ranging from editing chapters to asking loved ones to pick up life's slack in the big and little ways that family and friends do.

As for family, my debts are so deep that the best I can do is to dedicate this book to them and express my heartfelt gratitude. I want to start with an infinite thank you for my mom and dad, Sherry and Bob Johnson, who provided me, with an instinctive love and generosity, the type of support and security that made it worth risking time, money, and energies on a project and a career that promised few financial incentives but that meant so much otherwise. Thank you so much, Mom and Dad. After my parents, I want to thank my partner, Theresa Gallo, for the sacrifices she made to carry our family through the hard years of long distance, when a continent kept us apart, and then through absent-minded weekends when obsessive bouts of writing threw up new challenges. I love you and give this insufficient thank-you for carrying us through. Of course, I also want to thank my daughter, Ava. I'm so glad that you put your little muscular body between me and the computer—forcing yourself onto my lap and leading me by your little hand back to the park, the dinner table, and our family. Thank you for being alive, for breathing life into me, and for helping me to keep this book in its proper place. Beyond these, Kevin and Mike Johnson, Nikki Andrews, and the LaFayette and Gallo families have been supporters in both emotional and tangible ways throughout this process by loaning cars, footing dinners, being present, and doing the other things that generous people do.

As for my professional debts, I want to thank especially Alice Fahs, Ken Pomeranz, and Jon Wiener for steering me through the curves of academic life. Although I can't repay you for the many inspiring conversations I gained from having you in my life and for the endless letters of recommendations that you have written on my behalf, I can say a heartfelt thank you. I am so grateful to be able to count you as mentors, colleagues, and friends.

My other professional debts are more specific. Brian Black offered me great encouragement and criticism in reviewing this manuscript and in urging me to shoot for a bigger rather than a littler book. Colin Fisher was an excellent editor, friend, and sounding board throughout the writing process. Mike Maxwell offered the much-needed eye of an ecologist when I reached across the disciplines to a natural scientist to set matters straight. Joseph Mink and Miriam Wallace gave well-timed advice on the subject of coal, trauma, and literature when I first took up that subject. Fredric Quivik both pointed me to a few key resources early in this process and helped me to fact-check parts of the first chapter. Bob Lifset jumpstarted my work on petroleum's cultural history by sharing with me a list of historic films on oil that he had developed. Sandra and Bram Dijkstra were generous in donating a key image for this manuscript. Joel Dinerstein offered a precise and helpful critique of my manuscript in its first instantiation. And Mary Collar, Jim Diedrick, and Veronica Stewart planted the seeds of this book well before any words had been scribbled on the page. Moreover, because writing (to paraphrase Kenneth Burke) draws us into a conversation with those people who came before us and those to come after, I am thankful to the many scholars with whom I am in dialogue and whose names surface here and there throughout this text and its endnotes.

As for my institutional debts: While this book was written mostly during winter and summer breaks and in the early mornings, I received formal support from several institutions, including from my employer, National University, which offered two Presidential Awards to help me finish this book; from my former employer, the New College of Florida, which gave me summer funding and a half-year sabbatical to allow me time to complete some of the early research on this book; from the University of California–Santa Barbara for a Faculty Fellowship during which some of this book was first conceived; and from the University of California–Irvine, which funded through a Chancellor's and Dissertation Fellowship my earlier dissertation research, when some of the ideas in this manuscript were first developed. Moreover, because I have relied on librarians and archivists, I would like to say thank you to Matt Darby, who is the archivist of the Exxon-Mobil Historical Collection at the University of Texas–Austin for his help in directing me to the public relations material in that collection, and to the recently retired archivist David Klaassen, who years ago lent me his time as I worked through the

Survey Graphic collection at the Social Welfare Archives at the University of Minnesota–Minneapolis. And finally, I want to say thank you to Mike Briggs and the staff at the University Press of Kansas for seeing this manuscript through to the finish.

Thank you. This book wouldn't have happened without you all.
Bob Johnson

Introduction

Modernity's Basement

Philadelphia. May 10, 1876. Mud and scattered sun.

Americans stood expectantly on the edge of modernity. A sea of Victorian gowns and bonnets, black ties, and coattails. The occasion was scripted. At 1:20 p.m. in the afternoon, a Yankee businessman grabbed center stage to give the word, and on cue President Ulysses S. Grant and Brazil's Dom Pedro, emperor of the world's largest slave state, pulled a pair of twin levers to ignite the Corliss steam engine and to open the Centennial Exposition. Effortless action, so it seemed, jerked the Corliss to life, and a general clatter and thrum animated a mile of leather belts and steel shafts that vibrated throughout Machinery Hall. We are told that under the engine's steam power, a "thousand whirling emblems" of the nation's industrial present pulsed to life at that moment and that we, as a people, crossed over a historical threshold. Tears—wrote one journalist—fell from the eyes of a few strong men as they looked up and felt the import of the engine's curved flywheel and its elegant steel lines reaching 39 feet high to the rafters. But otherwise the nation's bourgeoisie simply burst into cheer and did not stop cheering for another hundred years.[1]

National memory falls along such well-trodden lines. The Corliss steam engine stands as a climax, a symbolic tipping point, in the collective consciousness of our industrial self. But the American emergence into modernity was never as scripted as that, nor was the break nearly so clean. In fact, the real event took place in a much smaller and less ornate building deliberately removed from the scene, a furnace room sequestered out of sight behind Machinery Hall. Here was an unceremonious space, the Corliss Boiler Building, a building named matter-of-factly for its utilitarian function. But in this furnace room—and in its analogs across the nation—a much deeper ecological revolution was taking shape and had been taking shape for a number of years. Like the stokehole of a steamliner and any number of other less prestigious boiler rooms throughout the country, the furnace room of the Corliss

The Corliss steam engine on display at the 1876 Centennial. Wood engraving, 1876. Courtesy of the Granger Collection, New York.

—which housed twenty upright boilers and a subterranean delivery system of wrought iron and firebrick capable of converting coal into the steam power of 1,400 horses[2]—was the endpoint of a breathtaking ecological paradigm shift that was changing forever the deep material relationships between people and their natural world.

Here, in this other space, in the absent center of our modernity, was being worked out a novel entanglement of rocks, bodies, and history. That entanglement would ultimately lift Americans out of a parsimonious world of limited organic resources—a world with fairly non-negotiable upper limits on material and population growth—and hurl them into a seemingly infinite new material world where the pleasures, risks, and dependencies of modern bodies could be tracked back to the lifelessness of mineral energies from beneath the world's soils and ocean floors.

Not all fairgoers made it to the boiler room. Not many of us do on a good day. But those who did got a glimpse of modernity's substructure —of the endpoint of a radical new ecology of production that had

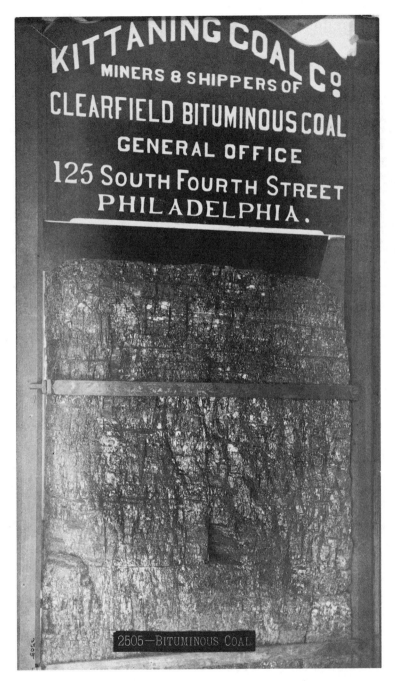

Bituminous coal: Kittaning Coal Company exhibit at Mineral Annex Building 46. Silver albumen print, 1876. Courtesy of the Free Library, Philadelphia, PA/Bridgeman Art Library.

fundamentally remade the material and cultural conditions of life under which modern Americans thereafter thought, circulated, and labored. It was here in the basement of the modern world, so to speak, that our energy flows, the first postulate of all life and human activity, were being shifted onto a subsoil ecology that anchored the modern self to the coal mines of Illinois, West Virginia, Ohio, and Pennsylvania.

Fairgoers—like those of us at the gas pump today—might have paused, as the fair's planners had hoped they would, to think about fossil fuels for a minute or two. Perhaps they stopped in Mineral Annex Building 46 to meditate on a clean-cut piece of bituminous coal mounted on backing board and encased in a wooden frame, or perhaps they admired the new coal-cutting technologies on display in a corner of Main Hall, but like most of us today, few of them fully understood what it meant to embrace these prehistoric carbons so eagerly on that day.[3] They too, that is, preferred to keep their ecological dependencies out of sight and to indulge instead, as the historian Andrew Goodheart has written, in the more anthropocentric fantasy of technology, progress, and the "perfect-running machine."[4]

Today, the costs of that neglect have become all too obvious to ignore. We have an increasingly pressing need to return to the origins of our modernity—to sort out the material substrata of our lives and to begin to make sense of what it has meant for us, in both ecological and ontological terms, to become a people of prehistoric carbon.

The Road Taken

1876 holds ceremonial import, but the actual year of transition was closer to 1885. The exact date is, of course, not of consequence, but it was in this later year, according to the US Department of Energy, that the scales tipped in the United States from the nation's historic reliance on renewable—and more clearly limited—energy resources like forests for fuel to its long-term dependency on subsoil carbons—coal, oil, and natural gas—to meet its rising energy needs.[5] In effect, two roads diverged in front of us in or about that year (one leading back into the forest and the other into the coal mine) and we chose to go down the one that had never been travelled.

Although that decision to follow down the path into infinity was only partially deliberate, the choice, if we can call it that, had enormous consequences. Within a matter of decades, coal, oil, and natural gas (once negligible features of the US ecosystem) were accounting for as much as 88 percent of the nation's total primary energy budget and were underwriting the nation's soaring population growth and wealth as each achieved undreamed-of heights. Nor were those dependencies transient ones. The depth of our commitment to fossil fuels remained remarkably unchallenged up through the present as those subsoil resources continued into the twenty-first century to constitute approximately 82 percent of the nation's energy budget.[6]

This book makes the case that we became a people of prehistoric carbon between 1885, when the United States experienced its first energy crisis, a "crisis of abundance," as historian Stephen Kern has called it,[7] and 1970, when we experienced a second, more depressing crisis of malaise. I argue that it was in these years—between the time Americans first broke through the barriers of the organic economy and nearly a century later when they seriously started to question the costs and long-term viability of doing so—that prehistoric carbons grafted themselves onto and embedded themselves deep within the American self.

Part I of this book, "Divergence," pursues that argument in two ways. First, it takes us into the basement of the modern world where this novel ecological arrangement was being worked out. By bringing us back to the coal mine and the stokehole, it reconstructs *modernity's ecology*—that is, the material logic by which this tidal wave of prehistoric carbon calories revolutionized Americans' access to energy and skirted us around the Malthusian constraints of the solar economy, both revising the nation's carrying capacity and wealth radically upward and generating a strong cultural aversion to the talk of ecological limits. Second, Part I traces out what we might think of as *modernity's ontology*—how this same flood of prehistoric carbon calories impressed itself in both conscious and unconscious ways on the modern body and mind, on our ways of being, knowing, and sensing in the world as Americans of different classes, races, genders, and conditions learned to embrace, absorb, and navigate the material manifestations and cultural potentialities of fossil fuels.

I argue here that becoming a people of prehistoric carbon in the United States meant learning to live with a distinctly modern set of ecological assumptions about wealth, growth, and well-being, and it meant absorbing modernity's subterranean energies into the body in multiple, if identifiable, ways.

Part II of this book, "Submergence," deals with the conflicted cultural record of Americans' reactions to these new mineral dependencies. Becoming a people of prehistoric carbon was not a passive event nor was it a straightforward one, and thus Part II, which takes its method and theory from Michel Foucault's work on countermemory and genealogy, recovers the cracks and fissures in the public discourse of energy and modernity in order to show how Americans registered, from more or less the beginning, a deep and sometimes tortured ambivalence toward their new path dependencies. Part II argues, in effect, that the nation's collective embrace of fossil fuels went hand in hand with both an unconscious and a deliberate suppression of those dependencies across social, spatial, symbolic, and psychic domains. By leaning on the work of cultural historians such as Peter Stallybrass and Allon White, this part of the book demonstrates that in order to live with fossil fuels, the nation's middle class learned to bury the human and environmental costs of its dependencies out of sight both geographically and socially in the working-class geographies of the coal mine and stokehole and to repress the many traumas and dislocations associated with fossil fuels both psychically and symbolically within a national narrative of progress, emancipation, and empowerment that had little room for modernity's objections and casualties.[8]

The main claim here is that this cultural work of suppression allowed for a modern bourgeois self to evolve that was untroubled by the collateral human and environmental damage caused by its dependencies on coal, oil, and national gas and dismissive of both its own traumas and those of the nation's many subaltern bodies, which experienced fossil fuels not as liberating but as disabling, constraining, and even atavistic.

The rest of this introduction provides the context for that argument.

The American Divergence

Americans have been mostly blithe about the role fossil fuels (or as J. R. McNeill puts it, "cheap energy") have played in their lives.[9] Although this neglect has been historically differentiated by region, class, and condition, and it arrived at different times for different Americans, the fact is that by the mid-twentieth century Americans as a whole lived, to quote historian David Nye, in the "most highly powered society in world history." The state's massive electrification projects in the 1930s, its investment in a federal infrastructure of highways in the 1950s, and its promotion of foreign and domestic energy policies that encouraged consumption worked to democratize access to modernity's energies in this period and to allow private energy interests to flood the market with prehistoric carbons while promoting an ever-rising consumption of energy to fuel the nation's multiplying automobiles, electric washers and dryers, refrigerators, and gas and electric climate-control systems. By midcentury, as Nye explains, the high-energy economy was simply second nature to almost all working Americans.[10]

As the beneficiaries of this unprecedented windfall of energy, Americans did not have to think very hard, as their ancestors once did, about the way that energy was accessed, the way it was processed, the way it was converted into work, the way it was recycled or lost, the ownership and control of it, the policies that governed it, the limits to its use, and most importantly, the ways in which it underwrites cultural and political capital. That type of intellectual work was left to the specialists. An abundance of energy—the per capita consumption of which in the United States climbed to the equivalent of about 23.6 barrels of oil a year per person or what translates—in terms of its power potential—into the manpower equivalent of about eighty-nine human bodies working for us day and night—removed us a step from the materiality of our world.[11] While Americans were, of course, reminded of that dependency every time oil prices spiked at the pump, when coal miners went on strike, and when petroleum hemorrhaged onto their beaches, public interest in prehistoric carbons typically receded very quickly into the background once the power grid came back on, once the strike was quelled, and once the well was capped.

The big unsettling fact about the modern relationship to energy is that our material dependencies came to run in the opposite direction of our cultural valuations of that energy—with our deepening dependencies on fossil fuels, as Wendell Berry once put it, perversely *cheapening* the value of that energy in our eyes.[12]

This seemingly benign neglect of fossil fuels has been, until very recently, especially evident in the disciplines of the humanities, which have been very slow to take up the subject of energy in more than a scattered way. Although cultural critics, of course, address energy on the sly in our disciplines (in speaking, say, of industrialization, machines, and coal strikes), and although we pay homage to fossil fuels in a short lecture on electricity and the White City or in reference to the railway in Leo Marx's *Machine in the Garden,* we have only recently opted to pull the subject out of the basement and confront it head on so that we might learn to develop the needed tools, the rich discursive textures, and the robust narratives for reckoning with what is the most fundamental material change in human history in the past eight thousand years. Scholars across the humanities have done that type of theoretical and conceptual work for categories such as race, class, and gender and for metahistorical topics like capitalism but not for energy, the one big thing that is arguably, as energy historian Vaclav Smil puts it, our world's most "truly universal" currency.[13]

The consequence of that reluctance is that while the topic of energy organizes entire disciplines and subfields in the natural sciences, those humanistic disciplines that take culture, meaning, and narrative as their main concerns (with the notable exceptions of ecological anthropology and environmental history) have not managed to integrate energy very effectively into the narratives of self and society that they use to frame their understanding of the world. We have important and pathbreaking works on which to draw, but the methodologies and narratives for centering energy in our studies are still evasive, and the scholarship remains piecemeal.

That lacuna is all too evident, for example, when we turn to historical narratives of modernity, modernism, and modernization that circulate in the humanities and social sciences. These narratives, especially in the work of cultural critics, have frequently sidestepped the epoch-making impact of this energy transition by seeking refuge in a sharp distinction between what is called the culture of modernism—which critics have used to classify a range of disparate events including the desacralization of nature, the rejection of Victorianism, the birth of nonrepresentational art, the ascent of relativism, and the advent of jazz—and the materiality of our modernity, or what we call modernization, which has typically been portrayed as the interplay of technology and economics. That strange tendency in the humanities to split the modern world between its cultural and material manifestations was once standard fare of cultural criticism, and it peaked sometime in the 1980s with Daniel J. Singal's claim in *A Companion to American Thought* that we needed to learn to "differentiate [the culture of modernism] sharply from modernization."[14]

The problem extends also, however, into scholarship we might reasonably expect to be more concretely grounded in the material world. For instance, even our scholarship on the history of technology, which brings us into a much closer engagement with modern material culture, still only skips lightly on the surface of modernity's ecology. Although it provides us with a rich understanding of modernity's energy technologies such as its dams, steam engines, dynamos, transmission lines, and electric gadgets, that scholarship opts for an almost exclusive focus on what we call power—our "consuming power," as Nye once phrased it—rather than the more elemental category of *energy flows* of which energy's conversion into power is only one part, albeit an important one. In short, even in our histories of technology (and it is also true of economics) we do not get a full reckoning with the material depths of our dependencies on prehistoric carbon. We leave that literature without knowing the one big thing that makes modernity possible—how this flood of prehistoric carbon calories redefined the ecological premises by which modernity's populations and wealth could soar to undreamed-of heights. To paraphrase environmental historian Ted Steinberg, while our scholarship on industrialization and its technologies has woven for us a bright tapestry of knowledge of the relations between humans and machines, it has not helped us to understand how the bustling and congested industrial world they built was rooted in a "tremendous ecological restructuring" that represented nothing less than a new chapter in the world's history.[15]

Part I of this book addresses that gap in our scholarship by following our dependencies on fossil fuels well beyond the more typical domains of technological developments, energy policy, and coal strikes to where they bear down on the deep ecological and ontological structures of modern life that shape the assumptions we carry around with us about what is normal, what is possible, what is sustainable, and what it means to live a good life.

This book is, of course, a partial contribution to that project, and it does not stand alone; thus the reader might turn to the endnotes for a record of the inspiring work on energy, ecology, and culture emerging at the interstices of the traditional disciplines, generative work that includes the writings of Stephanie Lemenager, Thomas Andrews, Anson Rabinbach, Andrew Nikiforuk, Stephen Kern, Richard White, Robert Vitalis, Brian Black, Myrna Santiago, Melanie McAlister, Edmund Russell, Carolyn de la Pena, Imre Szeman, Frederick Buell, and, of course, David Nye.[16] But here it is worth reminding ourselves that this growing insistence on energy's founding role in our modernity has its own roots in the writings of ecological anthropologist Leslie White, who argued nearly seventy years ago that Americans, like all humans, had passed through successive energy transitions, each of which has framed in basic material ways the possibilities, as he put it, of our cultural development. Today we do not have to accept White's notion that history is moving forward, nor do we have to suppose, as he seems to have, that materiality precedes culture, in order to accept the more basic fact that every society and every culture stands on its own bioenergetics, whether it draws its energy primarily from the world's grasslands, its coal mines, its oil fields, or its uranium deposits.[17]

Energy, that is, still sits at the bottom of it all, and thus we might find on this score some inspiration in the words of western historian Elliott West, who writes:

> We are surrounded by energy. It pours down on our heads and sits waiting in all organic matter, in tree bark, centipedes, cattle, bull snakes, and grass. Each life-form is constantly converting a tiny part of that abundance into an ability to do something. This is power's larger meaning. It is the translation of energy into act. It accounts for all true animation and allows every human event: a sacred dance, paddling a boat, burning a village, [or] sipping tea.[18]

The world's histories vibrate, that is, with nature's energies and their materialization into power and matter. From a culture's ability to perform a sacred dance to its destructive powers to burn a village, energy pulses throughout the social mechanism, and it is our access to these energy flows and our ability to capture, convert, control, and sustain those flows that serves as the foundation of everything else that we do as a species, a nation, or a culture.

Part I of this book might be seen, in other words, as an effort to restore to our narratives what ecologists and physicists have known all along but humanists somehow forgot—or sublimated—along the way.

Submergence, or the Sublimation of Fossil Fuels

The second claim of this manuscript is that while fossil fuels remade the ecological and ontological structures of modern life in the United States, Americans remained decidedly anxious about the dark underbelly of that bargain. I argue in Part II of this book, "Submergence," that despite everything fossil fuels such as coal, petroleum, and natural gas did to transform the material possibilities for economic and population growth in the United States, to raise national standards of living, and to generate the type of new somatic and psychic releases that came with extending the organic body's own limited capacities, they always tended as signifiers to steer Americans' attention away from the many joys and perquisites of modernity and back to its raw, traumatic, and repressed underside, where toxic spills, exploited workers, and injured bodies are the hidden costs of the country's growing energy dependencies.

The main story line in Part II is a cultural one that centers on the wars of representation that occurred between the nation's progressive critics—its writers, intellectuals, artists, filmmakers, politicians, and social reformers—who asked us to confront more candidly our disquieting dependencies on fossil fuels, and the nation's private (and sometimes public) energy industries, which aimed to steer us away from the mine and furnace and toward the world's brighter geographies where more sanguine histories of the modern self circulate untroubled. The method in this section is that of genealogy, which Michel Foucault defined as a type of unearthing of countermemory that seeks to restore to view the "unstable assemblage of faults, fissures, and heterogeneous layers" that get suppressed in our cultures as we go forward with the conscious and unconscious work of building and making sense of the worlds that we live in. Digging up that countermemory, Foucault tells us, means reckoning with the "heterogeneity" of the self and abandoning the sort of comforting notions of its stability and coherence that exist on the surface.[19]

For us that means returning to three types of historical sources: to the mostly forgotten work of big energy's progressive critics, who were frequently articulate about the psychic, social, and environmental costs of fossil fuels, to the tensions within mainstream culture's and even the state's representations of our dependencies on fossil fuels, and to the energy industry's own self-conscious efforts to manage our encounters with prehistoric carbon. In returning to those sources, we can start to recover a past that deconstructs the inexorable feeling that our modernity has been inching toward, to again steal a phrase from Foucault, some "monotonous finality"—a task all the more urgent because the finality we are approaching is an unacceptable one.[20]

Although this part of the book focuses on representations of fossil fuels and trauma, those representations cannot, as Part I makes clear, be viewed as independent of the actual traumas they sought to depict. The injuries suffered by Appalachian families living on the hard edges of the global mineral frontier[21] and the type of carcinogenic spills and decapitated mountaintops that continue to affect poor and even privileged communities today are just a couple of the examples of the range of social, physical, economic, and environmental traumas discussed in the historic literature that were quite real in any sense of the word. Recent events such as the deaths of twenty-nine coal miners in a Massey Energy mine or the public discovery that Kentucky coal miners regularly operate under the influence of hard drugs are merely the most current reminders that a systemic social trauma accompanies our world's scarred mineral frontiers and drives up the external costs of combusting carbon.[22]

These material traumas, which we see portrayed sometimes bluntly in progressive discourse, were, moreover, always partnered with a second set of less obvious psychic traumas that appear in the literature and imagery but that are more difficult to quantify. They include the emotional suffering evident in the grumbling of coal stokers, mill workers, slum dwellers, and even a neurasthenic bourgeoisie who were faced with the clamor, dirt, and disquieting pace of a fossil-fueled industrialization for which no premodern body could have been prepared. They extend to the stress and depression evident in the masculinist bravado of the exploited coal miner who has always made the best he can out of cycles of economic morbidity, personal risk, and structural paralysis. They are evident in the self-laceration that comes from the liberal bourgeoisie's fetishizing of preindustrial peoples (and other primitivist fantasies like the Tarzan phenomenon) that reach for psychic escape from the gesellschaft of their fossil-fueled world. And they can be felt in the alienation evident in the nation's compulsive repetition of pastoral fantasies that seek to return us to an agrarian ideal that never quite existed and that is no longer an option anymore anyway. That is to say, the perceptions of progressive critics in this book refer us back (albeit through a distorted lens) to a more concrete set of material and psychic traumas that go well beyond that of epiphenomena.

As for the term *trauma* in this book, it has not been used casually even if its usage does not always line up perfectly with the work of gender, race, and war scholars concerned with the injuries experienced by Holocaust survivors, postcolonial peoples, prisoners of war, and survivors of sexual assault. Etymologically, trauma denotes a wound to the body (or more recently to the mind), and thus the term's usage in this text is, in one sense, fairly obvious.[23] It does not take much to apprehend the explicit connections between the transition to fossil fuels and the type of material and mental wounds generated by that transition that were borne out in the bodies and minds of roughnecks, coal stokers, coal miners, factory workers, displaced peoples, and sometimes even a privileged middle class troubled by the environmental and social changes that the tapping of fossil fuels set in motion. So much seems uncontroversial, if no less interesting.

But the term *trauma* also carries with it a second set of theoretical and clinical connotations that refers not simply to a physical or psychic wound but to a type of physical or psychic wound harmful enough to a person (or, for our purposes, a collection of people) that it "bypasse[s] perception and consciousness," as Geoffrey Hartman tells us, "and falls directly into the unconscious."[24] In this other sense, the term *trauma* suggests something more complicated. It suggests a wound that because of its exceptional nature is difficult to process psychologically and therefore decidedly problematic to represent to oneself or to narrate successfully to others. (Oil's critics have, for instance, consistently made this point in arguing that our relationship to oil is so deep, uncomfortable, and even embarrassing that it defies narrative form and conscious recognition.)[25] This second meaning to the term *trauma* has some use-value for us, even if it is important to avoid conflating the wounds of a coal miner with those of a war victim or implying that environmental traumas such as the 2001 coal slurry spill in Inez, Kentucky, are in the same class as the more intentional assaults perpetrated in the name of race hatred.[26] But in each case, the term *trauma*—in this more graduated sense—seems to be the right one.

In this second sense, we can argue that the traumas associated with fossil fuels sunk—in at least an analogous way—below the register of the nation's collective consciousness, especially for its privileged and literate middle class. Because coal, oil, and to a lesser extent natural gas became associated with either damaged subaltern peoples and their peripheral geographies (such as the stereotyped "redneck" family in the far reaches of Appalachia or the migratory "roughneck" in the oil fields) or with the disavowed excesses of capitalism (such as those associated with the oil tycoon's unearned wealth and the industry's irrational booms and busts), they fell discursively outside of the normative values, images, and self-conceptions of the nation's middle class. That is to say that while fossil fuels came to sit squarely at the center of the nation's material core, they migrated paradoxically to the margins of its symbolic life, taking up residence not so much with the celebrated geographies of the modern city or with our glittering geographies of consumption (which, of course, depended on them) but instead with all of

Modernity's basement: Arthur Rothstein, *Stoking Furnace at Gopher Hole, Williamson County, Illinois*. Farm Security Administration, 1939. Office of War Information Photograph Collection. Courtesy of Library of Congress Prints and Photographs Division.

those other unwanted cultural objects that fall within the category of the "low"—symbolically devalued objects that although materially central to our lives are illicit and thus relegated to the basement of the world among the socially, psychically, and geographically marginalized.[27]

The consequence of that cultural devaluation of coal and oil is that they sunk—in a way decidedly out of step with their primary importance—to a sublimated role in the narratives we tell about the nation's coming of age and its unique historical trajectory. Whereas earlier cultures studied their energy flows intently while both celebrating and fearing their life-giving powers—such as the colonial farmer who cultivated with his own hands his farm's photosynthetic capacity as he converted the soil and sun's rays into the energy that his family could survive on—we industrial peoples have preferred to keep our energy dependencies out of sight.

Confronted, so to speak, with the gigantic Corliss steam engine at the 1876 World's Fair, we have typically seen in its power and in its hum the decidedly anthropocentric, and usually ethnocentric, story of how human innovation and an entrepreneurial spirit changed the world for the better. Like Henry Adams contemplating the "dynamo"—"that symbol of infinity"—we have tended to worship the machine while shoving downstairs the more humble and yet equally important ecological story about how we as a species learned to harvest new energy sources and reroute energy circuits to make that machine and the world on which it depends run.

This other more sobering story, which returns us to the much dirtier and worrisome world of finite resources and furnace rooms, does not fit very neatly into the tidy public spaces we have for celebrating energy and its power. We might say that in our public life, like our world's fairs, we have preferred, that is, to contemplate the fine steel lines of the Corliss steam engine in clean petticoats and top hats and let someone else do the worrying about its coal, soot, and slag.

That benign neglect has reached its own tipping point. It is now time to go back into the mine and stokehole to confront what we have repressed for a hundred years and more.

Carbon Nation

I | Divergence

1 | A People of Prehistoric Carbon

"To live for the first time on an accumulation of potential energy."
—Lewis Mumford, *Technics and Civilization* (1934)

Modernity's ecology arises on a mineral foundation. We are a people of prehistoric carbon deep into the bone.

That became true in the United States sometime between 1885, when Americans experienced their first energy crisis—a crisis of abundance[1]—and 1970 when a major oil spill and a series of oil shocks generated a second energy crisis—this time a crisis of malaise or confidence, to quote former president Jimmy Carter. During this period, the rise of the United States as a global powerhouse and the maturation of the nation's modernity were not simply (as we have at different times been told) the product of Yankee know-how, an amorphous Protestant work ethic, or even the nation's commitments to individualism, liberal markets, and competitive capitalism. Rather the birth of our modernity can be traced back to a breathtaking and often deceptively deep ecological revolution that occurred in the hidden structures of modern life and in the unspoken depths of the modern self—that is, to a new bioenergetics in our lives that occurred in the basement of the modern world where a firestorm of prehistoric carbon calories fueled the nation's economy, pushed Americans beyond the Malthusian constraints of the preindustrial world, revised upward the nation's carrying capacity, and remade the material basis on which American bodies thought, labored, and circulated in their world.

Modernity's ecology is palpable but it is not transparent. It takes a measure of patience to strip away the many layers of sedimentation that have accumulated over the years to distance us from the ecological core of our modernity. It takes some work, that is, to get into those deep subsurface structures—what historian Ted Steinberg calls the "basic stuff of life, the nitty-gritty of [our] biological existence"—that serve as the foundation of all of our other histories.[2] But by reacquainting ourselves with the ecological infrastructure that lies beneath our histories of technology and economics, we can start to understand why it is that Americans came to sit, at least for the time being, on the privileged side of history and how it was that the United States joined other industrial, and now postindustrial, nations high up in the box seats of modernity.

To that end, this chapter and the next address how prehistoric carbons came to presuppose the histories that we live: first, by turning to the role that they played in underwriting modernity's ecological preconditions; and second, by turning to the ontological structures that they fueled.

America's Culture of Big Energy

One does not have to be a crude determinist to know that modernity's many pleasures and hangovers, whether in the United States or elsewhere, hinge on fossil fuels—on the mine, the oil well, and the furnace. In fact, contemporary scholars after displaying a long-standing indifference to the material roots of the nation's prosperity have come around to this fact. Increasingly, they tell us that the nation's modernity, including not only its hard material infrastructure but also its patterns of being, thinking, and feeling in the world, can be traced back to this flood of prehistoric carbon calories. One critic, Frederick Buell, writes, for instance, that fossil fuels not only underwrite the nation's infrastructure of "roads, pipelines, telephone lines, [and] power cables" but also the habitus of modern life in the United States—our "private worlds, identities, bodies, thoughts, sense of geography, and emotions." And similarly, a growing number of other scholars including Stephanie Lemenager, Imre Szeman, Peter Hitchcock, Ross Barrett, and Daniel Worden have each claimed independently that fossil fuels are now so deeply encoded in our individual and collective lives as a form of "embodied" knowledge that they influence our unconscious pleasures, reason, senses, and emotions just as surely as they do our ability to commute to and from work.[3] We are coming around to the consensus, in other words, that fossil fuels are the absent center around which modern bodies and minds orbit.

Tapping fossil fuels was, of course, no small event. In contemporary parlance, it was what we call a game changer—the necessary, if not sufficient, factor for radicalizing and sustaining the nation's industrialization and for supporting the much larger cultural package of modernization that went with it. To name just the most obvious of fossil fuels' effects, they include: catapulting the nation's and the world's GDP to previously impossible heights;[4] unloosing unthinkable rates of population growth;[5] remaking the world's demography around dense new urban clusters and factories;[6] revolutionizing the nature and experience of work;[7] overhauling our perceptions of time, space, and even human subjectivity;[8] and reorienting people psychically around a new culture of growth that came to view progress for nearly a century and a half as though it no longer bore any relation to limited resources. Take away coal, oil, and natural gas—we would be living in a slower, rural, less populated, and more physically demanding world than the one in which we find ourselves. That is why earlier witnesses to this transition, such as novelist Sherwood Anderson, could claim so confidently that without fossil fuels the "great structure that expresses all the material progress of which we Americans are proud would be a useless thing."[9]

Of course, the United States never cornered the market on modernity. This world-historical transition to fossil fuels was a global rather than simply American phenomenon. But even so the United States stands out as *the* extreme case in that story. The facts on this score are relatively well known. Facilitated by a continent of largely untapped energy in the form of forests, farms, grasslands, and subsurface minerals, the United States made its great break with the premodern world in the middle decades of the nineteenth century by using the big energy in mineral-driven power technologies to exploit those resources with relative abandon. Taking its part in what world historians have called "the great divergence" or "the great transition" on the global scene,[10] the nation saw its production of power and its total energy consumption shoot up in this period at a rate unmatched by other countries—the latter of which first doubled every ten years between 1850 and 1910 during the height of industrialization (originally based on the exploitation of West Virginia's and Pennsylvania's coal beds and the rapid deforestation of the Midwest) and then increased another five-fold between 1910 and 1970 after the introduction of the even denser energy in petroleum from Pennsylvania, the Southern Plains, and California. The net result of transitioning to this subsoil subsidy was a 150-fold increase in Americans' energy consumption between 1850 and 1970 despite improvements in efficiencies—a fact that meant Americans could feel their energy consumption growing during each individual's lifetime for several generations running.[11] If we use the peak of the industrial age as a gauge of these changes, Americans were consuming something

The root of our modernity: Lewis Hine, *A Lump of Coal*, ca. 1912. Records of the National Child Labor Committee. Courtesy of Library of Congress Prints and Photographs Division.

like two-thirds of the world's petroleum, more than half of its coal, and nearly all of its natural gas—and thus far outstripping the rest of the world in its dependencies on prehistoric carbons.[12]

Like Great Britain, the United States never looked back once these mineral energies displaced the traditional calories culled from farm, field, and forest. Americans' embrace of what Lewis Mumford called "carboniferous capitalism" was eager, rapid, and seemingly irrevocable.[13] As early as the 1920s the average American could count on the support of an estimated thirty "energy slaves" per capita working for him or her day and night, or to put it in somatic terms, the mechanical equivalent of the thirty slaves, serfs, or wage workers that would have been needed under the preindustrial regime to generate the sort of labor power Americans were deriving from subsoil energies by that point.[14] Although the scope of that power only begins to tell the story of modernity's ecology, it is revealing of the changes under way because it was tangible, historically unique, and of a magnitude that previously only slavehold-ers, pharaohs, and monarchs commanded. Perhaps not surprisingly we too quickly came to take such power for granted.

It can be instructive to pause momentarily on the counterfactual—a Malthusian world without coal, oil, or natural gas. That world—heated by wood calories and dependent on food calories to fuel a biological workforce of men, women, and horses—would have permitted some form of industrialization to occur in the United States on a very small and localized scale wherever timber and food calories could be har-vested sustainably year after year and wherever reliable waterpower could have increased the nation's per capita productivity. But the more expansive phenomenon that we call modernity and the nation's partici-pation in it would have been a decidedly truncated, less ebullient—and probably more feasible—thing. We would have seen the Lowell textile mills replicated in a few reliable watersheds outside of New England, but industrialization as we came to know it would have been stillborn, its growth arrested by the fuel constraints of diminishing forests, by the modest potential of hydropower sites, and by the always-present biologi-cal limits that came with muscle power.

That is why scholars today suggest it is "impossible not to feel" that fossil fuels somehow structure our lives in a deep, unspoken, and subter-ranean way.[15]

An Orgy of Production and Reproduction

Among the competing definitions of modern life, two big material facts stand out that are not open for much debate. They concern what critic Lewis Mumford termed the "orgy of uncontrolled production and . . . re-production" that has defined the world since the coming of coal—that is, the sustained nonlinear economic and population growth that be-came possible after we shifted the world's aggregate energy supply and workload onto coal, petroleum, and natural gas.[16]

The figures are approximate, but we can get a pretty good sense of those changes from the writings of economic and environmental historians. What they tell us is that the global rate of both economic growth and population growth shot up to "spectacular" heights after we exchanged the old energy and power sources for fossilized carbon (and, in some places, hydro and atomic power). If premodern growth tended

Modernity arises from the coal mine. Detroit Publishing Company, ca. 1915–1925. Courtesy of Library of Congress Prints and Photographs Division.

to proceed worldwide at a snail-like pace, lilting along at a real growth rate that hovered just above zero[17] while adding only .6 new people per 1,000 every year to the planet (that is, more or less static wealth per person and a doubling of the population about every 1,100 years),[18] the coming of coal in western nations and Japan sustained new and unprecedented growth rates that climbed much higher than that while at the same time adding 6 new people per 1,000 to the planet each year—or what amounted to a doubling of the economy in industrial nations every generation or so and the world's population every 118 years.[19] Such modernized rates of growth, moreover, never ran into the type of Malthusian brakes one would have expected to hamstring the premodern economy when wood was a limiting factor and when food was the world's primary fuel source. Rather than a gradual slowing of growth as per capita resources diminished, what occurred was that growth continued unabated, albeit with a few sharp turns up and down, into the twentieth century, during which the world's gross domestic product (GDP) grew another fourteen times in size[20] (growing, for instance, in one generation between 1950 and 1973 by a mind-numbing 300 percent)[21] while at the same time adding, after 1950, as many as 18 new people per 1,000 per year (or a continued doubling of the economy and a doubling of the population every couple of generations).[22]

What is shocking when looked at from the perspective of the relatively "stationary" preindustrial world is that coal kick-started the world's economy so that it could grow forty times in size between 1820 and 1990,[23] so that its population could increase five times (from 1 billion to 5.3 billion) during that same period,[24] and so that it could experience real material gains per person in housing, clothing, food, fuel, and other consumer goods for a century and a half even as arable lands, pastures, and forests diminished per capita. The net result was that during those years we moderns, according to economists, came to acquire (on the average) about eight to ten times the real material wealth of our premodern counterparts.[25] If in an energy-poor preindustrial world rising populations and shrinking lands per person had always meant an inevitable decline in the standard of living, then something surely had happened. As economist Gregory Clark has phrased it, our "long sojourn in the Malthusian world" came to an abrupt end.[26]

The United States followed the pattern of other industrializing nations. Its case was, however, made unique by the fact that, unlike much of Europe and Asia, it still had a low land-to-person ratio and thus plenty of relatively untapped organic energy resources, or ecological slack, in the form of unfarmed land, unenclosed pastures, and uncut woodlands. That is, in contrast to the more populated Old World regions, the crunch of Malthus was only felt here and there in the United States and mostly along its eastern coast. As long as the violent removal of Native Americans proceeded apace, the nation's deep pockets of organic energy permitted demographic and economic expansion under the terms of the somatic regime.

The result in the United States was thus both similar to and different from the global pattern. What made the United States different was that it had already (even without coal) been expanding at an unusually rapid rate, "one of the highest in recorded history,"[27] because of the new ecological niches opened up by New World conquest. The figures are not perfect, but historians of early America tell us that the nation's rate of economic growth was remarkably high for a preindustrial economy.

According to John McCusker and Russell Menard, per capita real income grew at a rate somewhere between .3 and .6 percent annually in the colonial period, while gross national product (GNP) grew as much as 3.2 percent a year in the 120 years of colonization leading up to the American Revolution as European settlers integrated New World lands and forest resources into the global marketplace. Following the revolution, growth per capita might have climbed even as high as 1.3 percent annually until 1825.[28]

If the premodern economy was alive and kicking in America so too was the nation's population, which grew yearly on average by about 3 percent (or a doubling of the population about every twenty-three years) as the result of white migrants crossing the Atlantic to take advantage of New World resources and black migrants finding themselves chained unwillingly to that same task.[29] If per capita gains in real income were more or less static elsewhere around the globe, Americans saw something like a 50–100 percent rise in income between 1710 and 1840, although, of course, that income was very unevenly distributed.[30] Untapped land resources and thriving populations meant wealth in this preindustrial world, and thus free white Americans were consequently emerging as among the richest people in the world.

Yet two things distinguish US growth after fossil fuels from its premodern patterns of growth. While, on the one hand, the rate of annual growth in the GNP remained somewhat similar to that of the premodern world, skipping along at a very high rate of 3.6 percent annually for another 130 years after 1840, what did change dramatically was the rate of real gains per capita. The nation's historically high rates of real per capita growth before fossil fuels (made possible by the historic displacement of Native Americans) were still, in other words, "quite modest" by modern standards, the latter of which climbed during some years to five times that rate.[31] As Americans rode this wave of seemingly infinite energy into modernity, they saw real gains of about 1.9 percent a year for more than a century starting in 1870.[32] Such rates of real growth over such a long period of time translated into improved material standards of living that conditioned Americans to expect improvements over the course of their lives and that were historically unique.

Likewise, the size of the economy, which grew, as we said, by about 3.6 percent a year, continued to jump upward in the United States as though resources were not a consideration. That was true from 1840, when coal was first introduced into the nation's energy budget, through

1960, after which the nation's growth rates then peaked at about 4.5 percent annually during the prosperity decade before the oil crisis. During that long period of growth the US GNP jumped an astounding sixty-fold,[33] a remarkable pattern of growth that allowed the United States to emerge as the world's most formidable economic player and to acquire all of the perquisites that came with that power.

In terms of US population, the shift to coal did not initiate the sort of decisive spike in the rate of growth that demographers see in some other places, but it did ensure that what were already high rates of demographic growth could continue with little regard to the logic of the Malthusian economy. The point is an easy one to miss: because the United States converted early in its history to a new energy base of fossil fuels well before its people felt the pinch of the older organic limits, the nation could skip forward casually while absorbing Europe's surplus population and then parts of Latin America's and Asia's with little attention to the earlier ecological concerns (over such things as fuel and soil resources) that had always overdetermined the outer limits of growth in premodern societies. Although the *rate* of modern population growth gradually declined from its early breakneck pace when the continent was recovering from depopulation, hitting a high of 1.7 percent during the baby boom and a low of .7 percent a year during the 1930s Great Depression,[34] the nation's aggregate population nonetheless grew in spectacular fashion as it soared to nearly ten times its size between 1860, when we had 31 million people, and 2010, when we reached 308 million.[35]

What is so revealing about the American case from the long view, in other words, is that despite the pressures a relentlessly growing population placed on arable farmlands, pastures, and forests, the first of which shrank per capita from a potential 11.6 acres to 1.2 acres between 1860 and 2000, the second of which shrank from 18 to 2 acres during that period, and the last of which shrank from something like 25 acres a person to 2.7 acres per person, its people never quite felt the pain of that contraction in the way that their premodern predecessors would have.[36] Rather the supplemental calories of subsoil minerals made up for those losses in nature's bounty by propelling Americans (who still remained among the most well-endowed people in terms of organic soil resources) into what was the historically unique situation of realizing a steadily improving standard of living for several centuries running while watching farm and forest resources gradually diminish per capita and simultaneously managing to export large quantities of energy-intensive resources such as steel, wood, and fuel abroad in order to subsidize land-strapped regions across the Atlantic.

The important point is that the nation's early conversion—we might even want to say *premature* conversion—to fossil fuels, in effect, meant Americans unconsciously averted the Malthusian moment, evading the logic of organic constraints felt so viscerally in more land-strapped early modern regions such as England, France, Germany, Japan, and China.[37] Having inadvertently skipped around the energetic limits to the solar economy, Americans became subsequently vaccinated against talk of ecological constraints as they acquired what even today is a very strong cultural aversion to Malthus.

Some, of course, heeded the old warnings. Conservationists such as Pinchot worried that the nation's forests and mineral resources were being depleted at an unsustainable rate, eugenicists argued that the nation could not forever absorb throngs of new and undesired immigrants, war-wearied Americans knew about ecological limits from the sting of energy and other resource shortages, and contemporaries of the nation's baby boom and later oil crisis shared the sort of crisis in confidence that plagued all premodern societies. But such talk of limits either expressed itself primarily as periodic complaints about immigrants that had only academic attachment to ecological concerns over growth, or it found focus on the fringes of the national imaginary in the work of progressive environmentalists. Those concerns would not come to the foreground until much later in the 1960s and 1970s when we started to worry collectively about how our population boom and material profligacy were outstripping our available resources.[38]

To understand how Americans acquired their distaste for Malthus requires untangling modernity's ecology.

Modernity's Ecology, or the American Aversion to Malthus

John Muir once said that when we tug on one thing in nature we find it "bound fast by a thousand invisible cords that cannot be broken, to everything in the universe." The same might be said of fossilized carbon today. One can hardly walk, run, eat, or even dress oneself nowadays without having an encounter with fossil fuels. They fuel our cars; they ground our food supply; and they can be found on our bodies, shoes, sweaters, and nylons. We "ooze" petroleum is how one contemporary writer has put it,[39] although he probably should have included in that metaphor coal tar dripping from our bodies and natural gas emitting itself as well. But no matter how complete and complex our saturation in prehistoric carbon has become, we can still identify the core ecological functions they have performed since the great transition in the nineteenth century that allowed the world to spike demographically to 7 billion and elevate its economic output without "overshooting"—at least in the medium run—its so-called carrying capacity.[40]

Modernity's ecology, like all ecologies, is infinitely complex. That is true whether we are looking at the national case or the world-historical context. To understand it means reckoning with what Edmund Burke has called the "big story" of our species, that is, the causal relationship between fossil fuels and our species' long-term success both in multiplying itself and in enriching its productive capacity beyond the ecological limits of the world's previous energy regimes. It also means delving into what he calls the "bioenergetics" of modern life, or to put it less elegantly, explaining why our access to and use of fifty times the energy resources of hunters and gatherers and nine times that of advanced agriculturalists translates into growing populations, rising material wealth, and the impression of limitlessness.[41]

If we set aside for the moment concerns over climate change, which threatens to upend modernity in grand style, we can call out the four core energy shifts that made possible modernity's moment. Together, these changes that occurred in the basement of the modern world constitute the material premises behind both the nation's wild growth and its aversion to the more cautious logic of Malthus. We might see them as the postulates to our modernity.

To wit, they include:

1. the transmogrification of prehistoric carbon into energy slaves, that is, the conversion of subsoil minerals into the labor power needed to increase industrial productivity beyond what could be accomplished by food calories and muscles;
2. the impact that prehistoric carbon had on expanding the availability of the nation's heat budget for domestic and industrial purposes once actual forests were supplemented by what scholars visualize to be a "subterranean forest"[42] lying beneath the soil;

3. the use of hydrocarbons—coal tar, oil, and natural gas—as a material substitution for, or supplement to, soil-based fiber and wood products to meet Malthusian needs such as clothing and shelter; and

4. the conversion—through functional substitutions—of prehistoric carbon into the wheat, meat, and potatoes that we eat so that modern populations might increase without regard to the stingy logic of the organic economy.

Fossilized carbon thus means quite literally more of the Malthusian essentials of life for a domestic and global population that long ago exceeded its premodern carrying capacity. Fossil fuels don't simply translate into more fuel, in other words—they also allow for a set of ecological substitutions and functional equivalencies that stretch the earth's otherwise limited material capacity to produce the fuel, clothing, shelter, and food needed to support a modernized population. The trick to modernity's ecology is that it is so deeply encoded in this surfeit of cheap energy that we now hardly even notice the ecological weight those supplemental calories bear, although we come close to it when we hear writers such as Howard Odum lament that we are eating potatoes partially made out of oil.[43]

How coal, oil, and natural gas got us beyond Malthus is worth some comment.

Postulate 1: A Billion Unfed Horses, or the Transmogrification of Ancient Carbon into Energy Slaves

Fuel does not mean what it used to mean—or to put it more precisely, fuel does not mean *only* what it used to mean—that is, heat for stoves and blast furnaces and light for dark winter nights. Since the advent of the steam engine (1769), the dynamo (1831), and the internal combustion engine (1858), fuel is now regularly commuted not only into heat and light but also into motion or what we generically refer to as work or labor power. This transmogrification of minerals into mechanical energy had the historical effect of pushing productivity beyond the bounds of the somatic economy by unloosing what contemporaries imagined to be, in effect, the equivalent of 3 billion slaves into the nation's workforce, or the mechanical equivalent of 3 billion hardworking adult

The transmogrification of carbon into work: *Pres. Theodore Roosevelt on Steam-Powered Digging Machine during Construction of the Panama Canal.* H. C. White Company, New York, ca. 1908. Courtesy of Library of Congress Prints and Photographs Division.

bodies. By remaking work into a "mineral derivative," energy technologies, such as the steam engine and combustion engine, inadvertently allowed Americans to sidestep the logic of the organic economy and to enter into what has been (at least up until now) a more generous ecology of production.[44]

Man was the original prime mover—"the power plant that supplied the first cultural systems with their motive power."[45] That is how Leslie White was thinking about the subject in 1948. This equivalency that he saw between men and power plants had a certain ecological rationale to it. Prior to the steam engine, work was almost exclusively a biological phenomenon in which slaves, yeoman, and manual laborers moved alongside other beasts of burden like horses and mules. If today we talk about horsepower in the abstract, for most of human history literal horses along with men, women, camels, mules, oxen, and elephants were the world's most impressive energy technologies (as well as suppliers of fertility-rich manure and nitrogen). In their bodies, the grasslands of the dry West and the cornfields of the South could be turned into work, locomotion, and ultimately material wealth.[46] The rise of the horse-powered Sioux in the nineteenth century and the dense slave plantations of the West Indies are simply two of the more familiar instantiations of this biological energy regime, which drew its fuel from food calories and its power source from toiling bodies.

What defined the premodern world—and this was true regardless of time, space, and culture—is that its limits were bounded by the number of bodies that could be put to work or war at any given time during the year. Premodern cultures functioned, in other words, on what critic Andrew Nikiforuk has called "a two-cycle engine: the energy of solar-fed crops and the energy of slaves."[47] Biological labor was the basic motive force of that economy and food its most essential input. As classical economists had always stressed, living labor, whether male or female, horse or mule, young or old, was not free of ecological dependencies. It took a heavy toll on the environment. Even as early as the classical period, a place such as ancient Greece, which found itself forced to import food and fuel resources from distant ecosystems in order to sustain a large slave-labor force, was running up against the sort of strict Malthusian limits that came with amassing biological labor in quantity.[48]

The food requirements of biological labor alone pressed hard on the environment. A nineteenth-century American field slave or his yeoman counterpart demanded a dietary intake likely around 3,000 to 4,000 calories daily for an average adult male,[49] and those human motors would sap their own substance if their intake sunk below 1,500 calories even while not working. If we consider those calories from a purely technical perspective, they could be converted in the body of a healthy adult male into a meager one-sixth to one-tenth of a horsepower unit for the purposes of work. (To be even more precise, humans could convert food calories into work with an efficiency of about 20 percent, whereas other biological prime movers such as horses and oxen could convert calories into work at the much lower rate of 10 percent, making humans the logical ecological choice except when soil resources were slack or when soils would not produce more than fodder.)[50] Whatever the case, the expansion of biological labor pressed hard on the ecosystem with its relentless demand both for food and other soil-based resources such as timber for shelter, cordwood and charcoal for heating and cooking, and cotton and flax fibers for clothing. Relying on biological labor not only meant operating within certain technological limits (because unlike steam engines, which can work for long periods of time without maintenance, living bodies can consume and convert into work only so much food energy over so long a period of time), it meant reckoning any time that such labor expanded with what were increasingly heavy demands on forests, farms, and pastures.[51]

The key to long-term success or even short-term survival in these low-energy cultures thus always turned on stewarding available soil resources so as to provide biological prime movers with the material resources needed to keep them healthy enough to keep working. No American slave owner, for instance, could be casual about his land resources for very long. His ecological balance sheet always figured in the need to set aside either a significant portion of his plantation lands to meet the food, clothing, and shelter needs of a biological workforce or substantial profits from those lands so as to import these soil-based resources from someone else's ecosystem. The social landscape of the somatic economy thus tended to mean an immediacy of control over land, persons, and animals, and it tended to tie physical power to social power.

In the United States, the case was a little bit different. There was plenty of ecological slack to expand the nation's biological workforce under the older organic terms had Americans chosen to go down that path exclusively. Unlike more heavily populated regions in Europe and Asia, the United States had, that is, plenty of unexploited ecological niches to grow its population and economy under the old way of doing things without having to seek ecological relief through colonies, coal, or mechanized labor.

But even in land-rich North America, early Americans were not immune to somatic-era constraints, and the post-Columbian economy rarely expanded under the terms of sustainability. The repeated stories of Native American removal and western conquest are a reminder of the type of premium that the somatic economy placed on soil-based resources, which pushed Anglo-Americans to regularly coopt the forest, grass, and farmlands of Native American nations and Mexicans living on the periphery of their own ecological resources. Even if earlier Americans had chosen to operate sustainably under a low-energy strategy, the ecological constraints of that path can be extrapolated from the more fully developed Eurasian context, circa 1800. Here populations and economies (especially in the core regions of China and Western Europe) had already run up against the ecological limits to their growth under premodern terms, and were, as Kenneth Pomeranz puts it, in "serious trouble."[52] Their cases give us a sense of the counterfactual —where Americans would have ended up if they had not foisted their energy dependencies early onto subsoil minerals.

In such regions, more efficient usage of limited land resources (such as China's increasingly labor-intensive system of fertilizing, irrigating, and multicropping) and importation of land-intensive resources and goods (such as Britain's importation of Baltic lumber) could make up, to some degree, for the decline in per capita lands and forest resources that came with the growth of a biological workforce, but sooner or later the continued expansion of human prime movers in those regions ran into the inevitable environmental constraints imposed by limited forests, farms, and pastures. The shift to prehistoric coal calories in England for the purposes of both work and heat did, for instance, what local people, local wood, and local food calories could not; they dramatically expanded labor power and industrial productivity without further burdening the region's already stressed land resources. In the European case at large the way out lay in shifting to coal and acquiring distant colonies to add ecological slack to the system.[53]

What made the United States unique is that it adopted coal long before it felt the pinch of Malthus. Prehistoric carbon changed the equation in the United States before Mother Nature forced the issue. In quick succession, as economist Stuart Chase observed, carts pushed by hand gave way to mules and then to the locomotive.[54] This shift in aggregate terms from biological labor to energy, or mechanical, slaves represented not only a triumph of technology (as the story has typically been told) but also an ecological breakthrough. After all, different thermodynamic engines like men and machines put very different burdens on the environment.

In short, machines don't eat, and their expansion does not require the type of soil resources needed to shelter, fuel, and clothe a biological labor force. Their advantage from an ecological perspective is that they shift the burdens of labor off the soil and onto subsoil minerals. That difference, of course, revolutionized the bioenergetics of work and productivity in the modern world by lifting the ceiling on previous constraints.

Contemporaries estimated that by the end of the 1920s mechanical power was contributing more than 1.5 billion additional horsepower units into the nation's budget of available human power and literal horsepower, the latter of which had become in aggregate terms almost negligible.[55] That expansion of the nation's available labor force—seemingly without environmental cost—was the ecological predicate to large gains in net productivity and to the world's experiencing, for perhaps the first time in history, huge and continuing per capita gains in material wealth. The story no longer bears the luster it once did, but it is worth repeating in brief. The completion of the coal-fired transcontinental railroad in 1869, the shift to oil-fired tractors in the 1920s and 1930s, the development of the automobile and highway in the 1910s through the 1950s, and the proliferation of labor-saving devices fueled by mineral energy transferred much of the nation's labor burden onto mineral-powered energy technologies and elevated individual productivity in all major sectors of the economy. For example, in the nation's agricultural sector, man hours per acre of wheat plunged from 56 hours in 1800 to 2.9 in 1970; in cotton they fell from 185 to just 24 over that same period; and overall worker productivity in the United States increased every year for 130 years at a rate of 2.1 percent annually. One way to dramatize that fact is to imagine what it was like in the short period between 1940 and 1970 (when the high-energy society came to maturity) for the average farmer to put in the same number of hours but be five times as productive as his father or grandfather.[56] What that meant in the nation's once-dominant profession of farming was that mineral energy not only eliminated the need for much human labor, pushing people off the farm, but also terminated, in effect, a workforce of draft animals that had grown to about 16.5 million in 1920, or about .66 literal horsepower unit for every family.[57] This circumvention of biological constraints on labor in the agricultural sector was never the only factor (as we will see) in realizing the spectacular gains made in staple crop production such as wheat and corn, but it was one important piece of the environmental puzzle that pushed the country's yields to rise to spectacular heights well before the Green Revolution and the type of input-intensive agriculture that came to define yields after World War II.

From an ecological standpoint, as billions of metaphorical horsepower units poured into the economy, no one was wondering where those horses (or we might imagine human prime movers) were going to graze, eat, sleep, and dress. If we consider only the pressures on the ecosystem exerted by literal horses and mules in the United States circa 1920, when the actual living flesh in mule- and horsepower for transportation and agriculture was, as we have said, about 16.5 million and another 9.5 million nonworking mules and horses, we can start to glimpse the ecological costs to the premodern ecology of production. Feeding biological horsepower at its peak in 1920 required that a whopping one-quarter of the cultivated land in the United States be set aside in perpetuity simply for pastures and feed, and that was at a time when we had already advanced well away in our transportation sector from a dependency on draft animals to coal-fueled railways and automobiles.[58] In North America this ecological strategy of relying on horses worked for quite some time simply because 52 percent of US land, or about a billion acres, was usable for either pasture or cropland, which far outstripped the per capita soil resources of our European and Asian counterparts.[59] But to be sure, those draft animals were quite literally eating into the nation's potential food supply and occupying otherwise valuable ecological niches.

The ecological accounting does not end there. Our closest estimate from the period is that those 16.5 million literal horses, mules, and oxen were already being dwarfed in the 1920s by another 704 million mechanical horsepower units generated in the oil- and coal-fired engines of the nation's automobiles, freight trains, moving assembly lines, and cranes. Had Americans tried to convert such mechanical energy back into living labor, they would have been hard-pressed to come up with the foliage and pastures needed to feed something like three-quarters of a billion additional horses when Chase was writing or nearly 1.6 billion horses in 1930 after the nation's middle class had fully adopted the automobile. To imagine coming up with horse feed for more than a billion and a half horses, each of which would have consumed about 8.8 pounds of grain a day, would have required nearly 4 billion acres of dedicated cropland for draft animals, many times the available soil resources of the continent.[60] Thinking about horses in this way gets us closer to understanding the ecological cul de sac that defined the premodern economy and how coal offered a way out.

Agriculture was not the only sector of the economy revolutionized by resource energy. Mineral energy, of course, also broke through the biological and ecological barriers imposed on the nation's industrial and construction sectors, where fossil-fueled cranes, chainsaws, drills, assembly lines, die casts, and transport vehicles replaced the more limited capacity of men, women, and draft animals across the spectrum of activity. If waterwheels had expanded Americans' access to mechanical energy in places like the Lowell textile mills prior to the shift to coal by supplementing human labor with that of hydropower, the nationalizing of a mechanized workforce awaited the steady stream of coal calories that could detach production in all sectors of the economy from its earlier environmental conditions and biological constraints. Mineral energy, which contemporaries imagined to be the equivalent of "a billion wild horses," was the predicate, they said, to turning modern Americans into the type of "colossus straddling continents" we had become.[61]

Postulate 2: Averting Fuel Constraints, or the Subterranean Forest

The transmogrification of minerals into work was the most visible feature of modernity's ecology. But to be sure, only about 50 percent of this surfeit of mineral energy went into the production of power for industrial and domestic use. The other 50 percent of that energy, or what contemporaries estimated to be about 500 million tons of coal annually in the 1920s, was doing the same humble thing it had always done—generating not work but the big fires and heat needed to smelt metal, to liquefy silica, and to fuel the more modest stoves and furnaces of US homes. What one contemporary called the "art of applying heat" was as important to the nation's industrial takeoff as was the "art of applying motion."[62]

This revolutionizing of the nation's heat budget by supplementing wood-based charcoal with prehistoric carbons (which gave us the energy equivalent of another 686 million acres of sustainably harvested cordwood for homes and core industries such as steel, glass, and cement) represented a second ecological breakthrough.[63] That windfall of mineral energy, when taken together with the mineral energy needed for power production, pushed us past the preindustrial world's ecological threshold as it far exceeded as early as 1920 what the nation's actual forests would have been able to supply even at the dawn of colonization when virgin forests were widespread. That subterranean windfall of energy quickly became a structural assumption behind the nation's expanding wealth and population.

Energy as fuel for heat and light is a need of all peoples regardless of time or place. Premodern Americans, for their part, could be extravagant users of fuel, and wood was their energy of choice. The earliest Anglo-American settlers were not, after all, so unlike us. They too liked "warm houses," and they apparently also had an aesthetic penchant for big energy, in their case for "large fires." The average New England family might have used as much as thirty to forty cords of fuel wood annually, according to William Cronon's calculations, or one acre of forest clearance per year per family. Moreover, those thirty to forty cords of wood were in addition to the ecological burden colonists placed on forests for ship-, fence-, barrel-, and home-construction purposes. Although there is evidence that the more matured colonial economy in New England was thinking about fuel consumption and woodlots more carefully at the end of the eighteenth century, when forest clearance for farming was mostly completed along the East Coast, the net effect of such heavy energy usage was that the nation's fuel supply—its hickory, oak, and pine forests—was already being pressed upon hard in certain regions by the end of the colonial period.[64]

We hear of fuel shortages appearing as early as 1638 in coastal regions such as Boston and of colonists across the East Coast squabbling internally over limited resources in the commons while fighting externally with indigenous peoples for possession of coveted forests and soils outside of the colonial economy. To complicate matters, the nation's eastern forests were also serving a triple burden in providing deforested Caribbean sugar islands, voracious users of energy, with wood products and a tree-poor motherland, Great Britain, with the wood exports and wooden ships needed to subsidize a forest-poor metropole intent on floating the world's largest naval fleet. (The total timber exports by the early nineteenth century equated to about 1 million acres of wood annually from Canada and the United States.)[65] The overall picture is clear: although fuel shortages were not a continent-wide concern in the colonial period, we as a people were already drawing down our limited colonial inheritance of accumulated biomass and systematically cutting into the nation's preindustrial ecological foundation, albeit only regionally.

This pattern of drawing down our fuel resources only worsened in the nineteenth century after railroads, steamboats, and various power technologies placed new expectations on the nation's forests to produce fuel (as we have seen) for mechanical engines needed to run an industrial economy. Although we tend to forget the fact, firewood was the fuel of choice for steamboats and railways in the early stages of industrialization, and thus travelers crossing the United States encountered whole regions stripped of trees with large piles of wood sitting alongside riverbanks and railway stations for the purposes of feeding carbon-hungry steam engines.[66] Moreover, peopling the treeless plains and building an economy in the Great American Desert placed additional heavy demands on fuel and lumber imports from outside these energy-poor regions, which were less forgiving in their biomass and which were economically and demographically stunted until coal, as historian Thomas Andrews puts it, allowed them to "cross the industrial Rubicon."[67]

What that meant was that even before the United States underwent the type of full-throttle population and industrial growth that came to define its modern phase, all of the nation's major timber regions east of the Mississippi were already being called into service for fuel and construction purposes in an unsustainable way. That included, as environmental historians tell us, the broadleaf pine forests of the South, the white pines and oaks of the Midwest, and the white pines and northern hardwoods of the Northeast.[68]

To anyone who cared to notice, it was perfectly clear, in other words, that future population and industrial growth by firewood at the prevailing rate of consumption was going to be a losing game. However perverse it might be, looking back, we can see that it was this shift to coal that both saved what remained of the nation's forest lands and that can be credited for what Wallace Stegner once called "America's best idea," our nation's vast system of parks.

Modernity has been a Faustian bargain.

By the height of the nation's conversion to a fossil-fueled civilization, there were approximately 106 million people in the United States depending upon carbon-based fuels for their heating, lighting, and power demands. Absent coal, oil, and natural gas, the nation's 24 million households, each with about four persons,[69] would have needed something like two tons of coal equivalent (TCE) of fuel, or about three acres of woodlands per family, to be set aside in perpetuity simply to sustain even the low-energy demands of our early modern agrarian predecessors in the more straitened regions of France, Britain, and China. US forests (which stood at 754 million acres) could have easily met this limited fuel budget because that much acreage was still available in the United States, although transportation would have consumed much of that energy and setting aside forests for fuel wood would have put new pressures on harvesting wood for construction timber and for paper while also sending out timber exports to Japan and Britain.[70]

The situation, however, thickens ecologically when we take into consideration the fact that US fuel consumption by the dawn of the twentieth century was substantially greater, of course, than that of our parsimonious Eurasian predecessors and our profligate Puritan and Yankee counterparts; in fact, so much so that modernity's much-vaunted energy efficiencies would get us relatively nowhere. While early twentieth-century Americans might not have *visibly* burned large fires with the same abandon as their tree-rich predecessors did, they were still combusting per year (when industry is taken into account) approximately seven TCE per person (albeit fossilized carbon rather than wood) or the sustainable output of what would have been in the early modern period about a full ten acres of dedicated forest lands per person.[71] Coming up with that amount of cordwood, or about a billion acres of forest, would have been much harder. The last time North America saw that much forest was back in 1630 at the dawn of colonization.[72] And while much of that fuel went into the same basic purposes that it always had—from keeping American homes warm to blowing glass and melting iron for metal making—half of that fuel was going into the new thermodynamic machines that allowed for stamping metal, harvesting wheat, shipping containers, and achieving the economies of scale, enhanced productivity, and high outputs that characterize modern agriculture, transportation, and industry.

The point is simply that our dependencies on fossil fuels became so deep during this energy shift that we long ago passed the chance of reverting back to biofuels such as wood, ethanol, or bamboo to make up that demand—even under the most auspicious of terms. The figures are always loose, even sloppy, but by the end of the period under consideration, circa 1970, sociologist William Catton was estimating that the nation's entire corn crop of 6 billion bushels, if converted to grain alcohol, would have supplied just less than 2 percent of the nation's energy needs at a time when modern Americans were consuming approximately 58 barrels of oil per capita annually. Had Americans depended exclusively on soil-based fuels like wood or corn alcohol to keep themselves warm, to heat their stoves, and to run their engines, they would have found themselves as early as the 1920s with no remaining soil to grow food, fibers, or construction materials and a lot less remaining land to live on. The current fantasy of biofuels like ethanol thus is misguided in that it threatens, in effect, to reverse modernity's gains by placing the world's fuel burden back on the very same limited soil resources that we had for so long dreamed of escaping and that we still need for other purposes.[73]

To illustrate the ecological realities behind this fact, we might look briefly at how the shift from wood to prehistoric carbons affected traditional industries like the iron, brick, glass, and even salt industries, each of which was fuel intensive and each of which had previously depended on parsimonious woodlands. Historians tell us that such industries operated in the premodern period, except in rare circumstances, under the terms of fuel scarcity. Two examples can help to illustrate the point. In the case of iron production, the typical premodern foundry in a place like eighteenth-century Germany depended on large stretches of coppiced forests to produce the energy-dense charcoal needed to fuel its blast furnaces. A single ton (or 2,000 pounds) of wrought iron required the annual yield of a well-managed 24-acre coppice. Moreover, its bellows and forge hammers required that it be sited in a limited number of locations (i.e., near waterways) where mechanical energy was available for those purposes.[74] Consequently, many such foundries ran only for a few weeks out of the year and tended to be located far from population centers. To be sure, a ton of iron sounds like a lot, but it is not very much in modern terms. To put that number in perspective, the United States in 1920 produced approximately 33.5 *million* metric tons of pig iron alone.[75] Were we to try to produce that much iron by relying on charcoal and early modern furnaces, we would be stuck dreaming up another planet of forests simply to keep this one native industry in production. Thinking about the energetics of that one industry alone helps to explain why farmers and smelters were in a "thousand-year-old struggle" with each other for space on the soil.[76]

The trouble is that iron production was not, and is not, unique. Fuel (or the lack thereof) also dictated the output of each of the other major industrial sectors in the premodern economy that relied on heat, such as brick making, salt making, and glass making. Glass, which gives lucidity to modern life, was a particular "devourer of forests," requiring something like 2,400 pounds of wood to fabricate 1 pound of glass, albeit mostly for its potash rather than its fuel. Such wood was not easy to come by. In a preindustrial world prior to railways, the transportation of wood across land was labor intensive and expensive, which meant that manufacturers were dependent on a limited supply of fuel and materials from within a well-defined geographical orbit or from what could be shipped by sailboat from afar. In the previous case of German iron production, it was rarely economical or energy efficient, unless water transport was available, to transport charcoal beyond six miles. Consequently, the premodern output of the major industries such as metallurgy, masonry, and glassblowing was, by ecological necessity, always very small.[77]

What a cheap and seemingly endless supply of heat did was to revolutionize output in these industries while also, and as importantly, leaving behind plenty of additional fuel for the rapid and sustained expansion of other distinctly modern industries like the railroads, the automobile and trucking industry, and the various consumer goods industries, all of which were prodigious users of fuel as well. The way we might think about it is that the glass showcases of Wanamaker's in Philadelphia, the long futurist bands of steel girding the Brooklyn Bridge, and the white sheen of consumer durables in the modern kitchen all awaited this later radical materialization of cheap energy.

The upshot of it is that fossil fuels have since this early energy crisis played the part of the preindustrial forest, albeit many times over. By shifting to them in the late nineteenth century, we lifted our fuel pressures off of valuable land-based resources like woodlands while still managing to release a seemingly infinite stream of energy into the nation's economy without losing a beat. To visualize the material implications of this functional substitution of coal for trees, historian Rolf Sieferle suggests we might think of our coal and oil fields as an enormous "subterranean forest" underlying the soil, a vast prehistoric carbon inheritance much richer in energy than all of the living forests combined and so large that even today we can imagine using it to heat our furnaces and to drive our engines for another 200 years. Without that fossilized forest and its additional fuel calories, or some comparable substitute, we can be assured modernity's gains would quickly be erased.[78] This is why historians of the great transition can conclude with confidence that the "history of energy is the secret history of industrialization"—and it is why the heat potential in ancient carbon is a founding premise of the nation's modernity.[79]

Postulate 3: Averting Limits on Raw Materials, or the Curious Significance of Polyester Pants

We have thus far only examined the traditional stories of power and fuel, albeit from an ecological perspective. Economies and populations don't grow, however, without meeting each of the four Malthusian essentials of which fuel for heat is only one and power is only a means to acquiring them all. Fossil fuels did not only create ecological slack in our fuel supply and increase productivity per capita but also allowed modern Americans to loosen up the ecological conditions under which they could produce the other necessities of life like clothing, shelter, and food without overshooting in the medium term their natural resources. Without the ecological relief provided by that larger energy foundation, or some ready substitute, the world would be, to quote the late Franklin Delano Roosevelt, ill fed, ill clothed, and ill housed, poorer not simply out of social injustice (as much of it is today) but out of ecological necessity. This section briefly addresses why that is so with regard to Malthusian essentials such as fiber and construction materials.

1957 marked an important year. That was the year Americans walked into the world's first all-plastic home fabricated almost completely out of compounds of coal, oil, and natural gas. Ghostly white in appearance and with no natural referents, that dream home bespoke the nation's transition away from the old forest-borne materials of life, like oak and pine, and toward an increasingly synthetic environment fabricated in good part out of fossil fuels. A few years before that the nation's engineers had already rolled out imitation wood—a familiar-looking, if ersatz, version of the premodern world's built environment but one that was derived from oil-based polyethylene. And a third, symbolically pregnant, event from this time announced the additional importance of oil and gas to the nation's fiber supply. A small postwar riot in 1948 over petrol-based nylon stockings signaled our transition away from natural fibers for clothing to synthetic ones such as Fiber 66. Together these types of ecological subsidies—from synthetic fibers to hybrid construction materials—hint at the deeper environmental role fossil fuels played in stretching out modernity's material culture so as to meet the demands of a relentlessly expanding population.[80]

Despite the claim made by some economists that natural resources are cultural constructs, wealth has never been immaculately conceived, and natural resources, on some very basic level, still determine the possibilities for economic and demographic growth. There is, we might say, an ecology of production to all societies no matter how sublimated its material dependencies. In fact, as ecological anthropologists have shown, energy bears a tight relationship to the material standard of living, with energy and wealth serving as two sides of the same coin. While scholars have shown us that the human consumption of energy for endosomatic purposes has not changed very much over time (because the human body's capacity for consuming food calories can only vary so much—by about two calories for the hungriest of us to one for the most malnourished), they have also helped us to understand that the exosomatic use of energy, that is, the energy that goes into our material culture, has expanded by leaps and bounds since the coming of coal. Architectural historian Luis Fernández-Galiano, for instance, claims that the prodigious amount of energy that modernity has accreted in its built environment in the form of materials such as glass, concrete, steel, aluminum, and plastic is a defining feature of modern life.[81]

In such a context we have to take with a grain of salt the comments of earlier social observers such as Buckminster Fuller, who wrote that modern life was becoming increasingly "ethereal" because of its abstraction from the material world. The truth of the matter is that we have always been running in the opposite direction—toward a materialism so intense that many of us in resource-rich nations like the United States no longer worry about basic material matters such as where our food is coming from as much as we worry about the risks of living our lives measured not by quality but by a materialist gauge contemporaries referred to as the "quantification" of life.[82]

The modern economy, despite its many energy efficiencies,[83] is predicated on an enormous per capita consumption of material resources that calls us back to the thirty cords of real wood that we saw previously being hauled out of the forest annually by our energetic Puritan predecessors. Stuart Chase calculated what it would mean in material terms to pile up on the lawn of a suburban family of five the "flood in raw materials" such a family depended upon by the 1920s. That family would, he said, have to own a very large lawn for its:

25 tons of coal;
6 tons of quarried stone;
1,475 board feet of lumber;
2.7 tons of iron ore;
317 bricks;
38 barrels of oil;
7 barrels of cement;
638 pounds of salt;
180 pounds of sulfur;
98 pounds of copper;
67 pounds of lead;
40 pounds of rubber;

. . . [and the] smaller heaps and piles of aluminum, silver, zinc, raw silk, cottonseed cake, nitrates, [and] antimony." That is to say, even before one started to consider the "tons of raw foodstuffs" needed to support an American middle-class family in the 1920s, it would have been no simple task to try to dig that family out from underneath its mountain of material dependencies.[84] (Today, per capita consumption of raw materials in the United States amounts to about 46,000 pounds of material resources annually—or the equivalent in weight annually, if you will, of a 23-ton sperm whale.)[85]

To understand how coal, oil, and natural gas relaxed constraints on the production of essential materials for housing, furniture, and clothing, it is instructive to look briefly at the nation's material culture prior to the coming of coal in the United States.

What is clear when we look at preindustrial material culture is that it derived from what we colloquially refer to as organic compounds. Clothing, for its part, derived from cotton, flax, wool, fur, leather, and sometimes silk, and the premodern built environment, especially in the United States, primarily derived from wood—from oak, maple, pine, and hickory—although adobe, brick, and stone played their parts as well. If we leave aside the question of clothing, wood left its heavy "imprint," as historian Joachim Radkau puts it, on the "material-sensuous nature" of nearly every society up through the nineteenth century.[86] Although American merchants and planters could sometimes afford to import or produce fuel-intensive bricks baked in kilns for their homes and buildings, although stone was a viable material in some regions where labor was plentiful, although glass played its part in the material culture of early America, although adobe was critical in fuel-scarce western regions, and although iron and ceramics were available for purposes such as stoves, nails, tableware, and luxury goods, it was wood, especially in the East and Midwest, rather than these more energy-intensive materials, that dominated the everyday built environment.

The list is long. The vast majority of the nation's homes and public buildings was made chiefly out of wood, or in some regions energy-intensive bricks and labor-intensive stone, and wood was critical in almost all cases to roof beams, floors, doorjambs, windows, and interiors. Early American tools such as rakes, oxcarts, barrels, looms, gears, spinning wheels, and even sometimes pipes were made out of wood; transportation vehicles such as ships, carriages, and wagons were fabricated out of wood; power technologies, such as wind and waterwheels, were made mostly from wood. And the trappings of early Americans' domestic environments, including their beds, chairs, tables, and cradles, were constructed out of wood. The forest was, in other words, a one-stop department store before we actually had department stores.[87] Because Americans benefited from having colonized a continent of relatively uncut woodlands, their built environment and their material culture arose out of the forest. It is worth remembering that there was a time in early America when metal products were rare enough that a merchant could corner the market on iron nails and a time when fuel- and labor-intensive masonry was a privilege of the well off.

If premodern America's built environment required the forest, so too did its clothing fibers derive from soil-based resources. King Cotton, as is well known, competed for most of the lower South's soil resources from the beginning of the nineteenth century up through the early twentieth century, during which Great Britain's and then the nation's own booming textile industries first took off. Likewise flax, wool, and leather each directly or indirectly derived from the soil, from pastures, woodlands, or farms. What that meant from an ecological standpoint is that before coal, oil, and gas made other fiber and building materials accessible and cheap, the nation's built environment, including its clothing supply and material goods, derived almost exclusively from the soil—from timber-rich oak forests, fiber-rich cotton farms, and woolly sheep pastures. In this premodern environment, the nation's fiber materials and its housing materials were, in effect, crowding each other out for space on the same soil that was needed to produce other essentials like fuel and food.

Modernity did not change that fact completely. The nation's love of wood and organic textiles continued relatively unabated. The annual production of timber for nonfuel needs, such as housing, paper, and exports, stood at about 75 cubic feet per person annually in 1920, and the pressures that textile production put on the soil included another 35 million acres of southern farmland set aside exclusively for the cotton crop, or what amounts today to about 9 percent of the nation's operating farmland.[88] But what is unique about modernity's ecology is that the land available for producing that wood and those fibers was available precisely because the nation's forests and farms were no longer being asked to produce cordwood and hay to fuel the economy and to meet the needs of an army of preindustrial prime movers. That is to say, there came to be in this period a large, if indeterminate, number of what Catton calls "fossil-fuel acres"—real land acreage that was freed up from its historic usage as fuel and pasture and repurposed for the production of building materials, textiles, and human food crops.[89] As just one example, the fact that members of the US middle class could, up through the twentieth century, still purchase with relative economy a wood floor for their homes testifies to the beneficial impact fossil fuels had on relieving pressures on woodlands for the production of timber. The big point is that these emancipated acres became a hidden premise of modernity's material culture once they were no longer needed for their traditional tasks.

Fossil-fuel acreage meant that more land could be put aside for timber and cotton. But the impact that coal, oil, and natural gas had on expanding the resources needed for fabricating modern life went well beyond that. By revolutionizing the nation's heat budget for primary industries like metallurgy, cement making, and plastics, prehistoric carbon also allowed for the mass production of nonorganic materials for the first time in history. A "wall of steel" rose up in the modern United States as the nation divorced its material culture from the forest and rebuilt itself around new compounds composed of subsoil ores, rocks, and fuels.[90]

Energy was, of course, the critical factor in facilitating this substitution of timber with metals and other energy-intensive materials like cement. The smelting needed to produce steel and aluminum; the quarrying, baking, and transporting of cement; the spreading of ribbons of asphalt; and the proliferation of nonorganic materials like plastic represented nothing less than the material manifestations of fossilized carbon. Without such cheap energy, steel and concrete would be spare rather than ubiquitous, and plastics, with perhaps the exception of products such as cotton-based cellulose, would be of minor importance. This substitution of other engineered materials for wood allowed the nation's material culture to take shape under wholly different—and more relaxed—environmental conditions than it did in the nation's earlier era of wood.

The most striking ecological change in this novel material culture concerned, of course, not the expansion of the old materials of life, such as steel and concrete, but the shift to these other wholly new synthetic materials composed of coal tar, natural gas, and petroleum, materials such as polyvinyl chloride (PVC) that did not exist prior to the shift to fossil fuels and that extended the nation's ecological foundation into

Congealed carbons and the making of a world of steel, glass, and concrete: Alfred Palmer, *Molten Iron Is Blown in an Eastern Bessemer Converter to Change It to Steel at the Republic Steel Corporation, Youngstown, Ohio,* 1941. Office of War Information Photograph Collection. Courtesy of Library of Congress Prints and Photographs Division.

uncharted terrain. Synthetic substitutes like plastics, which ultimately came to constitute about 4.6 percent of the nation's total annual oil consumption and 1.5 percent of its natural gas consumption, moved the country another step away from the premodern world's forests, sheep pastures, and cotton farms.[91]

This shift from pure timber to synthetics was most evident in the nation's built environment, where the ubiquitousness of plywood in construction, a compound of fossil-fueled resins and wood invented in 1934, represented by midcentury the hybrid nature of the nation's material culture wherein the products of the soil were supplemented with and/or replaced by subsoil carbon to meet Malthusian ends.[92] Not only did plywood and imitation wood give to postwar housing its characteristic flexibility and plasticity, these and other hybrid materials stretched the resources available for the mass production of homes. Similarly, the use of petroleum-based asphalt starting in the late nineteenth century allowed for a radical expansion of the nation's transportation infrastructure as oil helped to build 41,000 miles of new highway following the 1956 National Interstate and Highways Defense Act.[93] The case was less dramatic but also relevant with regard to clothing and other nonhousing and noninfrastructural materials, since it was also in this period that synthetic fibers such as nylon and polyester, which derived from coal tar, gas, and petroleum, started to compete with organic textiles to define the intimate texture of modernity's clothing, bedding, and drapery.

Norman Mailer captured something of the ecological significance behind this shift to synthetics. In 1963, he wrote that oil-derived plastics had divorced "humanity from its past" and propelled us "into a fundamentally nonhuman artificial reality." Although plywood, asphalt, and polyester pants might lack the "odor of the living," as he put it, they also came to underwrite, for better and for worse, the nation's wealth and its ecology of production. We might look at imitation wood and imagine that we see the real thing, but in ecological terms we are looking at a very different sort of reality that arises on a very different set of ecological assumptions.[94] Plastic might not be unreal but the reality it signals is not the reality we once knew.

Postulate 4: Averting Food Constraints, or the
Promises of Synthetic Bacon

Of course, none of these developments would have mattered much
without the crops needed to feed a rapidly growing population at home
and abroad. Perhaps the most important thing subsoil energies did was
to expand the nation's and world's food supply in the twentieth century beyond the limits of the organic economy. Although it might be
counterintuitive to associate fossil fuels with ecological generosity, the
fact is that prehistoric carbon has both freed up arable lands for the
purposes of food production and allowed agriculture to intensify in a
way previously impossible under the somatic economy with its heavy
burdens on the soil. This hardwired connection between fossil fuels and
the world's food supply became by the height of this first energy shift
a deeply embedded premise of modern life that structures the assumptions that Americans make about the realistic limits to growth. Fossil
fuels reshaped, in other words, the size of the edible pie that was around
to share as the nation's population expanded.

The day that we as a species dreamed of converting coal into "synthetic bacon" marked an important turning point. That event, which occurred in 1929, acknowledged in effect just how deeply dependent our
food supply had become on fossil fuels. Although it would, of course,
have been simpler to eat coal or drink petroleum directly, very few people—with the exception of a handful of German scientists, apparently
some pregnant women, and eight trapped Chinese miners—have tried
to do just that. Not all energy is equal, and the body is ill-equipped to
turn ancient carbon directly into food calories.[95] But through a series
of ecological substitutions, modern Americans did learn to eat potatoes, wheat, and bacon that were made at least partly out of coal, oil,
and natural gas. The reasons are multiple but we can adduce two basic
changes in the energetics of modern agriculture that explain modernity's bumper crop and its subsequent population boom beyond the
warnings of Malthus.

The first has already been treated to some extent but requires elabo-
ration. As we have seen, one of the big material impacts of fossil fuels
was to take pressure off of land resources so that they might be reallocated to other purposes. Catton's concept of "fossil-fuel acreage" ap-
plies here as well as it points us to the fact that coal, oil, and natural
gas emancipated land otherwise needed for purposes other than food

production. Although it is hard to get a perfectly clear picture of how much land was actually freed up and then used in the United States or elsewhere for food production by this shift to fossil fuels, we can trace out with confidence the logic that was in play by returning briefly to the functional substitutions described above in which subsurface coal and its successors took the place of soil-based resources like trees, fibers, and construction materials for the purposes of sustaining a modern-size population. As just one indication, by the 1920s the estimates were that approximately 310 million acres of forest lands (which might have once been needed for fuel) were more or less permanently converted to food and textile production.[96]

This story of the forest is already familiar. Whether we consider the nation's population of 106 million in 1920 or its 203 million in 1970, the point is that its population had been freed, beginning sometime around 1885, from having to set aside woodlands to meet its fuel needs. The dependability of the subterranean forest ensured a steadily rising supply of fuel without the need to set aside real forests to satisfy the energy demands of a constantly growing population. (Of course, if Americans had continued to rely on the forest, even a country as large as the United States would have had to have been by the 1920s carpeted in pine, bamboo, and corn for ethanol simply to meet the nation's fuel needs, making it a theoretically inhospitable place to live with little food to go around.)

Although the estimates we have are simply ballpark figures, they can serve to dramatize the point. Were Americans to rely on the sustainable production of, for instance, an early modern forest to meet their fuel needs, they would have been tasked by 1970 — given their heavy energy dependencies — with coming up with about 15 acres of forestlands per person simply for fuel production.[97] Even in a land-rich country like the United States, we have only 2.7 acres of forestlands per person and 1.2 acres of farmland per person to meet our individual soil demands; and for better or worse almost all of that land was already reallocated over the course of the twentieth century to producing timber for nonfuel purposes and to being cleared for growing the fibers and food crops for a population that has sextupled since 1885. The big picture is that during this first energy transition the nation's forestlands were repurposed from producing fuel to other tasks such as producing construction timber or being razed for food. Fossil fuels allowed some of Georgia's and North Carolina's broadleaf pine forests to be put into the production of peanuts, pigs, and chickens; Michigan's, Wisconsin's, and Minnesota's white pine and oak forests to be shifted over to producing corn, soy, and hay; and Pennsylvania's and New York's former forests to produce hay, corn, fruit, and soy. The remaining 700 million acres of forestlands in the national domain that we still have today were, moreover, set aside, as we have said, either to produce raw materials for the nation's timber and paper needs or to provide modern Americans with a measure of psychic and physical relief from the city.

The story of the pasture and meadow (where the nation's draft animals once went to refuel themselves) is similar. The shift to coal, oil, and natural gas for power production, as we know, rendered draft animals obsolete. Because in the United States feeding horses, mules, and oxen was often done on land that could otherwise be used for human tillage, the switch to tractors and trains (which ate coal rather than grass and hay) freed up by sometime in the 1920s and 1930s approximately one-quarter of the nation's cultivatable land for the purposes of food production. By terminating a workforce of 16.5 million hungry draft animals, in other words, the national soil bank saw a large bump in its potential acreage for human crop production. Subsequently, many of those liberated acres were called upon for use after World War I and during World War II, when former pastures and meadowlands, including substandard soils, were either plowed under for the purposes of food and fiber crops or repurposed for stock animals rather than workhorses. Those decades saw the logic of extensive agriculture reach its modern peak in the United States sometime around 1950, when aggregate crop yields skyrocketed on the treeless and horseless farm.[98]

This disappearance of draft animals from the farm did not, of course, mean a disappearance of all animals from US soil. In fact, Americans came to raise on that emancipated soil a large annual crop of edible animals, which today has climbed to upwards of 34 million steers, nearly 9 billion broiler chickens,[99] 110 million hogs, and 2 million sheep and lambs,[100] plus another several million dairy cows and egg-producing hens that don't go straight to the slaughterhouse. But those animals were by midcentury repurposed for food production rather than for work, and thus their life cycles, their biology, and their grazing habits were re-engineered to maximize the number of edible calories they could yield rather than the work calories Americans could get out of them. If at one point a good workhorse grazed on hay, fallowed lands, and pastures while supplying up to twenty years of service for the family farm, those bucolic creatures were replaced over time by the modern world's "power steer" and its "Mr. McDonald" chickens, animals the likes of which have since the late 1950s been pumped with estrogen and asked to live out a life cycle that has almost nothing to do with their locomotive capacity and everything to do with their edibility. Today, the typical steer lives out its life within an abbreviated fourteen- to sixteen-month life span during which it is expected to emerge from the womb, to grow from 80 to 1,200 pounds, to show up at the market, and to be consumed so as to free up room on the land for more of its kind.[101] This transition from the family workhorse and pasture to modernity's revolving power steer is an illustration of the modern ecology of food production that became possible after Americans abandoned draft animals and reclaimed the resources they once required. Were Americans to have shifted their labor burdens back to work animals, there would be a lot less meat to go around, no room left to grow fibers, very little grain, and a whole continent of alfalfa, hay, corn, and oats to feed hungry, and sometimes ornery, mules and horses.

Fossil fuels did not, however, simply free up soil resources but also allowed for an intensifying usage of those soils. Up through the 1940s the main story concerned the extension of agriculture into new lands freed up from other purposes; but equally important is that fossil fuels permitted a radical *intensification* of agriculture after the logic of extension had played itself out. Agricultural gains were increasingly made in the postwar years not by moving into new ecological niches opened up by, among other things, fossil fuels but by intensifying energy inputs into lands that were already being farmed. An estimated 12 percent of the nation's total energy budget in the 1960s was going directly into production activities on the farm or indirectly into the inputs and transportation sector needed to support these modernized farms.[102]

During this period of intensification, coal, oil, natural gas, and hydroelectricity along with, to some extent, nuclear power intervened in the nation's and world's food supply in several direct ways. Perhaps most importantly, big energy (of which fossil fuels are its predicate) was the essential resource needed for the synthesis of artificial fertilizers that allowed farmers to break from the natural nitrogen cycle and other mineral uptake cycles so as to keep soils in production year in and year out. The high pressures and high temperatures required to fabricate ammonia urea and other nitrogenous fertilizers assumed a constant input of cheap energy starting sometime after World War I, without which farms would have been constrained to the much slower process of fallowing lands and keeping them out of production for long periods of time until they could recover.

Perhaps it does not sound like much, but 1 percent of the world's total energy consumption now goes directly into producing the heat needed to capture and turn atmospheric nitrogen into the 100 million metric tons of ammonia, urea, and urea ammonium nitrate applied yearly to the world's soured farmlands to feed a population that exceeds by about 40 percent the capacity of the premodern nitrogen cycle.[103] Likewise, the birth of large-scale phosphate-mining operations in places such as Florida during the 1920s—which dwarfed premodern exchanges like the early modern guano (i.e., nitrogen) trade between Peru and North America[104]—and the mining of potash or potassium from ancient seabeds starting in the 1960s allowed energy-intensive extraction processes to replace the older and much more constrained business of fallowing lands and supplementing those lands with both organic nitrogen sources and potash from charred trees.

This proved to be the crucial input for the success of modern hybrid crops, like the F-1 hybrids, that went into production after World War II and that brought yields in staple crops such as corn, rice, wheat, and soy to stratospheric heights. If an acre of arable land yielded about 20 bushels of corn in 1920, that same acre was yielding by the end of the century as many as 180 to 200 bushels an acre, a rise in calories per acre that permitted each American farmer to support the caloric needs of about 129 people.[105] Prolific and yet fragile, those hybridized crops depended not only on petrol-synthesized fertilizers to guarantee high yields but also on an energy-intensive water infrastructure and a protective net of oil-based pesticides, herbicides, and fungicides to prevent infestation of fields that lacked the natural genetic diversity to defend themselves. Forty percent of energy use in the nation's agriculture was going by the end of the century into fabricating the 5.1 billion pounds of petroleum-based fertilizers and pesticides that shield those plants.[106] Less obvious but equally important is the role cheap energy played in pumping and moving water cross-country to supply drought-intolerant hybrids with the uninterrupted irrigation they demand. The nation's massive infrastructure of pipelines, aqueducts, and pumps that grew up after the 1902 Newlands Reclamation Act altered the nature of food production by relying on cheap energy to move water across deserts and mountains and to drill and pump deep underground aquifers.

There is a great irony to all of this. Although modernizing the food supply made US agriculture vastly more productive by midcentury, although it was a great saver of human energy, and although it allowed the nation's and world's populations to soar to new heights, its energetics was primitive from the start. Overall, the modern world spends about five times more energy producing its food than did hunters and gatherers and about twice as much as did traditional farmers before the start of industrialization.[107] In the United States, Americans today use nearly three times the energy to produce the same amount of food as that of their neighbors in developing nations.[108] What that means in general is that we spend something like seven calories of energy (mainly from fossil fuels) in order to produce one calorie of food that we eat.[109] To put it a bit differently, we burn about five pounds of oil, or a gallon of oil, in order to produce a pound of farmed salmon.[110] The payoff in doing so is that we save prodigious amounts of human energy in our farming, at least the energy of most Americans, but only by substituting subsoil mineral inputs for that labor. Such an abysmal level of energy efficiency in our food supply would not have been tolerated by the premodern world's more energy-efficient farmers, but it is tolerated by modern farmers because the calories that go into producing modernity's food supply derive from a previously useless part of the ecosystem (i.e., subsoil minerals) and do not use up the more precious human labor and edible energy from the soil.

Working through the logic of modernity's ecology in this way helps us to understand what Howard Odum meant when he said Americans were eating potatoes "made of oil." The benefit of shifting the nation's food dependencies onto prehistoric carbon was, of course, obvious. It meant that the nation's and world's population could shoot up beyond the ecological limits of organic agriculture. But the risk was that Americans found themselves hurled into what McNeill calls a "gigantic uncontrolled experiment" in which they were "gambling" that those high-energy inputs could be sustained in perpetuity to allow them to operate outside of an organic paradigm that had been more or less intact for the previous 10,000 years.[111]

A Modern America without Malthus

The American adoption of big energy was the critical factor in its sharp break with the premodern world and its equally sharp cultural antipathy to the warnings of Malthusians. The historic and mostly hidden ecological subsidies that big energy provided by the late nineteenth century made it, in other words, very easy for modern Americans, all the way up through their second energy crisis in the early 1970s, to skip along with little notice of the type of energy constraints that had just previously shaped the world of their predecessors. Even today, with 97 percent of the world's scientists in agreement that fossil fuels are causing irreversible changes to the earth's climate, the language of "drill, baby, drill" still resonates on the political right while the center and the left talk about "growing" ourselves out of our problems. The presence of big energy made it seem natural, that is, for a land-rich nation like the United States, which was already cashing in on its natural resources, to jump into modernity's orgy of production and reproduction without regard to the sting of organic limits. It was not simply, as the story has often been told, the nation's work ethic, its eager embrace of capitalism, or even its Yankee inventiveness that explain the characteristic American brand of optimism that has accompanied the nation's material success at home and on the global stage but rather these subterranean subsidies that by 1885 were propping up a radical new ecology of production that had resulted in a booming economy, a rising population, and a strident aversion to the rest of the world's concern with limits.

2 | Rocks and Bodies

To accomplish the work done annually in the United States . . .
would require the labor of three billion hard-working slaves.
  —Chester Gilbert and Joseph E. Pogue, *Power* (1918)

Three billion hard-working slaves, or approximately thirty slaves, serfs, or wage workers for every man, woman, and child in the United States.[1]

That is the number of healthy adult bodies Americans would have needed—according to two mineral specialists at the Smithsonian—to convert their daily consumption of mechanical energy back into living labor at the peak of industrialization. In the years leading up to that figure, they explained, mineral-driven technologies—such as trolleys, tractors, trains, cars, compression drills, water pumps, harvesters, and conveyor belts—had multiplied in the United States and were regularly converting coal and oil into something like 150 to 430 million horsepower units for a population that hovered around a hundred million people. Based on their assumption that it took about twenty healthy adult males to generate one horsepower unit, the Smithsonian's curators calculated that the average American was thus being propped up by the work equivalent of something like thirty "slaves" or "servants" per person—albeit fossil-fueled slaves that, as they put it, "knew nothing of the wear and tear and hopelessness of a servile life."[2]

This equation of minerals to enslaved bodies was not just a disarming analogy. It reflected the fact that fossil fuels had destabilized on a basic somatic level both Americans' access to work and the modern body's relationship to its material world. By drawing a direct correlation between our growing dependency on these new fossil-fueled slaves and our ancestors' more literal dependency on biological slaves, the Smithsonian's curators were, in a sense, grasping to find a way to narrate and to normalize the dramatic dislocations that came with transitioning from a low-energy world of hooves, feet, and hands to a new and unfamiliar one wherein work was increasingly disembodied—or, as they put, a "mineral derivative."[3] Leaving aside for the moment the uncomfortable racial unconscious of that analogy, we might see this figurative embodiment of fossil fuels as a sign that the most elemental relationships among energy, work, and the body were being recalibrated and that the deepest grounds of our being were starting to tremble and shake.

We might take it as a sign, that is, that Americans were starting to see, feel, and think in a distinctly modern way and that modernity's ontology was coming to take its shape.

Modernity's Ontology and the Role of Prehistoric Carbons

The body, critics tell us, is porous, situated, and constructed. It is a historical thing—produced by and responsive to specific cultural and material contexts.[4] That being the case, we might ask how modern bodies (and our ways of knowing those bodies) were rebuilt around the combustion of prehistoric carbons and their consumption as power, heat, and light.

To date, there is very little literature on this historic relationship between energy and the modern body—how our senses, our emotions, and our basic experiences of our own corporeality in the domains of work and leisure were reconditioned and redefined by the infusion of natural gas, petroleum, coal, and other modern fuel sources into our lives. Recent work on this subject has, however, begun to take hold in the humanities, where critics increasingly talk about our nation's current "oil ontology"—that is, the ways in which our being, feeling, and sensing in the modern world are structured around petroleum and its products.[5] Such literature, although it tends to overstate the importance of petroleum by ignoring the persistent role of oil's prehistoric companions coal and natural gas, has nonetheless moved us in the right direction by joining the history of fossil fuels to a theoretical literature on the body, which claims that the corporeal and lived self is a situated and environmentally disciplined, rather than universal and autonomous, thing.

A particularly striking example of this mode of thought comes from critic Stephanie Lemenager, who writes that modern bodies are now so thoroughly embedded in an infrastructure of petroleum that such intimate matters as our "muscle memory" and our affective relationships to the material world are conditioned by, or at least wrapped up in, petroleum flows.[6] To phrase that a bit differently, we might say that we "body forth" a self rooted in prehistoric carbon every time we step on the pedal, turn on our bedroom lights, melt into a warm shower, or take off from the tarmac.[7] We are a people of prehistoric carbon deep into our nerves and muscles (albeit in socially differentiated ways), and thus there is no extracting the contours of the modern self from an external environment saturated in coal, oil, and natural gas that shapes its experiences. We are oily, yes, but also sludgy, tarry, and gassy deep into the bone.

This chapter teases out the main facets of modernity's ontology by identifying how the shift to prehistoric carbons affected modern bodies (and our experience of those bodies) in two related ways. First, it addresses the *disembodiment of labor* from human agents by demonstrating how this historical transition to mechanized horsepower—in aggregate terms—compelled us to rethink and repopulate, in some very basic ways, the concepts of energy and work around a new thermodynamic discourse that collapsed what were previously hard distinctions between food and coal and people and machines. Second, it turns to the complementary *embodiment of fossil fuels* that went hand in hand with this development by tracing out how modernity's energies were absorbed into, appended to, and apprehended by these newly modernized bodies, each of which came to know themselves and the world of prehistoric carbon in socially differentiated ways—including at the points of carbon's production, at the points of its combustion, and through the bodies' corporeal encounters with an exosomatic environment made out of the congealed energies in coal, oil, and natural gas. Together, these related developments constitute the structures through which fossil fuels were scripted onto modern bodies and the ways in which they rewrote our ways of knowing and being in those bodies.

Modernity's Energy and the Disembodiment of Work

Fossil fuels had sweeping impacts on the modern body. Historically, their most dramatic impact, however, turned on the role they played in destabilizing the categories of energy and work—two of the most basic features of our biological life and among the most elementary building blocks of all historical societies. Entering a thermodynamic age—that is, an age in which impassive minerals and active human bodies were both doing, and sometimes competing with each other to do, the world's labor—required a radical overhauling of these two basic concepts, especially as they bore down on the status of laboring (and leisured) bodies in modern societies.

First, energy. The multiplication of thermodynamic engines in the mid-nineteenth century revolutionized the meaning of energy. The invention of the steam engine—which converted the heat potential in carbon into the equivalent of working muscles—forced us to rethink the definition of energy around a new thermodynamic paradigm that centered on the heat and work potential of minerals, food, engines, and muscles. Prior to that development, Americans, like all westerners, had used the term, which derived from Greek and referred to *vigor of expression* or *a life force within something*, in what today we would consider a figurative manner.[8] Poet Phillis Wheatley wrote, for instance, of the "glowing energy of thought" the god Apollo bestowed on writers; Thomas Jefferson wrote about the "value of energy" in human governments; and philosopher Ralph Waldo Emerson wrote that there was a "unifying . . . energy" in our lives that amounted to an oversoul connecting us all. Nowhere was energy yet a materialized object, and it had very little to do with physical forces.

The steam engine, in a sense, coaxed us into a more materialist definition of energy that pivoted not on transcendental forces but on the ability of nature's varied resources (including its rice, grass, coal, and uranium) to function as fuels in the bodies of both animate and inanimate prime movers for the purposes of work. What we inherited from that moment was a brand-new conceptual order that gathered nature's diverse objects—from its edible resources such as rice to its mineral resources such as bituminous coal—under a new discursive umbrella, where they were reclassified as energies (or fuel resources) and at the same time subjected to a new measurement of calories that denoted their heat, and, by extension, work potential. That mental reshuffling of nature around a thermodynamic paradigm introduced us into a wild new cultural order wherein we could now imagine that the meat in a thousand-pound Angus steer was the energy equivalent of approximately eighty-eight pounds of coal—a very weird way of thinking and talking about the natural world that would have sounded strange to our forebears.[9]

The cultural consensus came to be that energy—no matter what physical form it took—represented *the capacity to do work*—a novel way of thinking about the subject that placed food in the same category as petroleum and that turned muscles and machines into analogs of each other. Although energy, of course, continued to carry its humanistic connotations deep into modernity, there came to be this sneaking suspicion that when we were being serious about the subject, we were not meditating in the yoga studio or praying to the gods but had our feet firmly implanted in the domain of physics, where lumps of uranium and organic chickens, human bodies and metallic engines, stood alongside one another.

Second, work. Reconceptualizing energy went hand in glove with rescripting the meaning of work. Prior to the steam engine, work had almost always referred to the physical activity of human bodies and to the labor of horses, mules, and oxen. In that preindustrial context, a handful of nonbiological prime movers, such as windmills, watermills, and sails, could provide Americans with a degree of mechanical energy to augment what was available to them in muscle power, but otherwise such mechanical work was scarce, and machines could only here and there be said to be the equivalent of human muscles. The defining feature of this premodern world was, in other words, that it was quantitatively and qualitatively a *somatic* one anchored in working bodies and in what we might call the muscular economy.

The advent of the steam engine challenged the primacy of this muscular economy by expanding the definition of work to include the work done by thermodynamic engines. Although the transition was piecemeal, mechanical energy seems to have begun to outcompete the nation's biological workforce of 16 million horses—at least in its sheer provision of brute horsepower if not precision—sometime around 1890. Thereafter, both the ratio of mechanical work to living labor widened and the changing ontological status of rocks, bodies, and machines only deepened.[10]

Increasingly, as Anson Rabinbach has shown, physiologists, physicists, and engineers could see manifested all around them new functional equivalencies between the "human motor"[11] and the inanimate engine, both of which were rechristened prime movers during this transition. The advent of the term *labor power*, for instance, and the newly debiologized concept of *horsepower*—both of which emerged from this moment—further collapsed what had once been vital distinctions between living beings and their technologies. From that point forward, we could start to measure the world's different types of work, whether done by humans, mules, or machines, through a new and standardized calculus of energy inputs and work outputs, with the consequence that work lost its uniquely human and bodily dimensions.[12] The change was so rapid, Rabinbach tells us, that even such a strident proponent of living labor as Karl Marx came in this period to rethink the meaning of work as he abandoned his initial claim that work was a form of self-realization (he called humans *Homo faber* in his early writings) to rethink work as a burden on the body from which fossil fuels and the new energy technologies might emancipate us.[13] Of course, the climax to that ontological shift arrived with Frederick Winslow Taylor's attempts to render the human worker more energy efficient, and thus more machine-like, him- or herself.[14]

Contemporaries, for their part, saw this disembodiment of work to be the defining feature of their modernity, and they commented on it both regularly and with great zeal. A documentary such as Pare Lorentz's 1938 *The River*, a state-funded film about the promises of hydro-electricity, can exemplify for us the sort of stark contrasts contemporaries drew in this period between the once-dominant muscular regime that was becoming socially and economically marginalized and modernity's new high-energy and power-rich geographies that were defined by electric motors, electric appliances, and an assortment of gas-powered backhoes, cranes, tractors, automobiles, and harvesters. Two scenes from that documentary—each of which focus the camera in on the hands, arms, and thighs of straining bodies at work—help to evoke for us the texture of this earlier muscular economy that the nation's middle class was deep in the process of forgetting.

In the first scene, Lorentz turns the lens on a team of working mules dragging mud and dirt from the Mississippi's shorelines up to its embankments. Here we see the silhouettes of mules and mule drivers move across the screen not once, not twice, but three and four times to capture both the intensity and the repetitiveness of the raw physical tasks to which these working bodies have been put. As bulging thigh muscles and hardened hooves come into close frame, the camera draws the viewer into a brief intimacy with these living prime movers as they toil and strain to pull their heavy load of rock and sludge up a muddy slope. In a subsequent scene, the vantage point shifts from these draft animals to human bodies. This second somatic sequence shows a group of African American sharecroppers moving step by step in real time as they harvest what appears to be an interminable cotton crop in the South's hot fields. Here the corporeal burden rests not on the sturdy thighs and backs of draft animals but on these hunched and nameless workers whose muscles seem to be grooved to the sort of preindustrial

Embodied work: John Vachon, untitled, 1938. Office of War Information Photograph Collection. Courtesy of Library of Congress Prints and Photographs Division.

rhythms that once characterized most American lives. "Men and mules," the narrator says in tired repetition. "Mules and mud a thousand miles up the Mississippi"—the phrase capturing in word and rhythm the slow cadence, the repetition, and the muscle-aching tempo of this earlier somatic workscape, which beat to the pulse of hoof and foot.[15]

We can glean from a film like Lorentz's the texture and grain of a preindustrial world that defined the experiences of labor—albeit in socially differentiated ways—before high-powered transmission lines, pounding dynamos, and whirling turbines swept the nation's middle class into an amped-up world of electric farms, homes, cities, and offices. That is, we can get a feel in an artifact like this one—even if that feel is ideologically driven—for the type of preindustrial sensory regimes, muscle memories, and erotic textures that once stood at the center of a preindustrial nation before fossil fuels started doing much of the heavy lifting for its more privileged classes.

But in order to get a more concrete gauge of the nation's transition away from the muscular economy, we might also follow out the plunging physical exertion of the average American body over time. All the way up through about 1870 the majority of Americans had worked out of doors on farms and plantations and burned something like 1,000 calories a day in the fields while consuming about 3,000 food calories total. Over the course of the next few generations, however, many of their children, their grandchildren, and their great-grandchildren moved indoors, purchased mechanical servants, and sat by watching as their physical exertion plummeted to something like 300 calories a day, or what was less than one-third of the muscular output of their recent ancestors.[16] By the peak of this transition, economists were counting as many as fifty-four different types of hard physical labor that mechanical servants were doing in the stead of actual human bodies and draft animals. Pounding, crushing, blowing, sucking, pushing, pulling, digging, hurling, sawing, lifting, twisting, swimming, boring, drawing, weaving—the list they came up with was nothing short of a list of human kinesthetics.[17]

A second imperfect measure of these changing relationships between the body and labor can be intuited from the staggering increases in our reliance on disembodied power for once common tasks such as hauling wood, pumping water, sowing fields, and harvesting wheat. If mechanized labor—with the exception of watermills and windmills—was mostly negligible before 1800, only a few generations later, circa 1880, Americans were depending on 14 million steam-powered horses to supplement what was available to them in living labor, most of which was lodged in the hauling and transport capacity of their coal-fueled locomotives. That figure only climbed thereafter. It had jumped to 45 million horsepower units by 1900 (for a population of 76 million), once the nation's railways expanded and steam-powered factories restructured its major energy-intensive industries. And then by the 1920s and 1930s, following electrification and the widespread adoption of the combustion engine, Americans could claim an almost incomprehensible figure of 1.6 billion mechanical horsepower units in the form of gas-fueled tractors, automobiles, trucks, buses, and various electric motors. That sort of disembodied power dwarfed, of course, what they could muster up through their own bodies and what was available to them in the 17.6 million literal horses and mules that populated the US landscape in 1930.[18]

If we know the world and ourselves through labor—"we know nature through labor" is how Richard White puts it[19]—then this disembodiment of work that fossil fuels introduced into our lives had deep ontological implications that are worth teasing out.

But before doing so, a caveat is in order. Modernity's energies were not evenly distributed. They did not produce a singular experience, nor can we say that our encounters with them were even mostly benign. Sweaty and demanding physical labor intensified (as we will see in the later sections of this chapter) for subaltern bodies in the nation's new high-energy workscapes, and it persisted in a traditional muscular economy that never quite went away in professions such as agriculture and construction, even if the nation's farming population fell from nearly 90 percent to just 3 percent between the nation's founding and the late twentieth century. Modernity's privileges and its pains were from the start, in other words—like its predecessor's—socially stratified, and frequently so along familiar race, class, gender, and regional lines. The bone-deep structural changes that came with embracing fossil fuels have, that is, a social history to them.

Moreover, the disembodiment of labor generated, even among the nation's middle classes, almost as many lamentations as it did celebrations. In fact, just as the relative weight of living labor was in decline, a new sentimentality for the muscular economy emerged—as horses, mules, and servants quickly resurfaced as objects of romance and endearment in American culture. Such sentimentality can be heard in Teddy Roosevelt's loud exhortations to middle-class men to recommit themselves to the muscular conquest of nature and the world's "primitive" races by taking up the "cult of the strenuous life." It can be sensed in the sudden appearance of bourgeois fitness clubs on the modern landscape, a late nineteenth-century development popularized in New York City. It can be seen in the nation's Works Progress Administration (WPA) murals, which eroticized thick male necks and the full forearms of the nation's producing classes. And it can be felt in Americans' nostalgia for an embodied, and even racialized, somatic work regime in films such as the 1939 Best Picture, *Gone with the Wind.*

That is to say, Americans might have eagerly embraced modernity's energies, but we felt a certain collective ambivalence about that embrace.

Big Energy at the Point of Consumption, or the Birth of a Modern Prosthetic Body

Of course, the body never went away. Tapping big energy was not simply a matter of disembodiment. It also produced distinctly new bodies

The modern romance of men and their muscles: Carl Morris, *Agriculture* (WPA mural, 1942). Gary Halvorson, Oregon State Archives.

that worked differently, that circulated in new ways, and that had a different understanding of their own corporeality, including the sensory texture of the world around them. We can start to see the contours of our modern ontology by outlining the four main modalities by which modernity's energies were subsequently absorbed into, appended to, and apprehended by modernity's different bodies in socially stratified ways.

First, the corollary to this disembodiment of work was the birth of what critics have seen to be the modern prosthetic body,[20] a liberated and predominately middle-class body whose senses and psyche were reconditioned to a heavy dependency on fossil fuels at the point of their consumption. This complementary embodiment of fossil fuels experienced by modernity's privileged subjects was at the core of modernity's claims to being emancipatory since it had made possible the type

of amped-up and exhilarated bourgeois body that today we take for granted. As theorists of technology have long known, a sort of collective cathartic release was part and parcel of modernity's emancipation from the traditional somatic constraints of muscle power since fossil fuels, to quote one critic, "facilitate[d] the body's capacity for self-extension toward *other* life."[21] Although modernity's new energies might have been "ambivalent," to quote Lewis Mumford—with their "depressing effects contrasted sharply with their stimulating ones"—there is little doubt that fossil fuels liberated the modern middle-class body from previous physical constraints and ushered in a wave of exhilarating new affective and sensory experiences that sunk deep into the body, giving shape to a modern eroticism that even today keeps us committed to fossil fuels.[22]

We can glimpse this experience of energy as bodily and psychic extension in the ways that aviator Charles Lindbergh talked about his experiences flying. Lindbergh, who worried that the comforts of modernity were "insulating man from life," also spoke of being shot through with "adrenaline" and being most fully alive physically and mentally when he was in the cockpit of his gas-fueled *Spirit of St. Louis*. Here is how Lindbergh described his experience as a prosthetic human being:

> While my hand is on the stick, my feet on the rudder, and my eyes on the compass, [my] consciousness, like a winged messenger, goes out to visit the waves below, testing the warmth of water, the speed of wind, the thickness of intervening clouds. It goes north to the glacial coasts of Greenland, over the horizon to the edge of dawn, ahead to Ireland, England, and the continent of Europe, away through space to the moon and stars, always returning, unwillingly, to the mortal duty of seeing that the limbs and muscles have attended their routine while it was gone.[23]

Lindbergh's juxtaposition of the corporeal experience of commandeering the new energy technologies with the psychic release he felt in controlling a scope of energy that carried one beyond the organic body could hardly be stated more explicitly. By drawing a direct link between the physical experience of channeling a material power previously found in the realm of fantasy and the altered consciousness that came with the mastery of that power, Lindbergh experienced the now-familiar spatial and temporal liberation that critic Stephen Kern has argued fossil fuels made possible.[24] The underlying sensuality, or even eroticism, in his experiences—the thickness of moving air, the hurrying of the wind, the imagined warmth of the water below, the expansiveness of his vision, and the feel of the plane's seat, stick, and rudder—all point to a self thoroughly embedded in the experience of prehistoric carbon. We can come to understand why Lindbergh, as a prosthetic human being, spoke about his plane in the first-person plural as "we," man and fossil-fueled machine, and why he could say that he in a sense "loved" that plane with a certain sensual attachment.[25]

Flying a biplane across the ocean was not a representative experience. Not everyone had access to that sort of prosthetic freedom. But some more pedestrian version of those psychic sensations and erotic attachments to the new energy technologies certainly occurred across the spectrum of modern life as Americans reorganized their muscle memory around the power, the touch, the smell, the sight, and the sounds of automobiles, washing machines, airplanes, and tractors. This embrace of the textures of big energy as a means to physical and psychic liberation was based on the very real fact that the modern body, or at least modern middle-class bodies, was increasingly free of the constraints of

Charles Lindbergh and the prosthetic body in flight, 1930. Courtesy of San Diego History Center.

an organic body that had once been limited in its geographic mobility and tied to a routine of physical burdens such as hauling loads of wood by foot, lifting heavy irons by hand, pumping water by hand, and tilling dirt by back and plow. The ontological restructuring that attended our becoming a people of prehistoric carbon centered, in other words, on this retooling of the modern body's conscious and unconscious experiences around the acquisition of a power that reached well beyond the scope of human muscles.

The primary symbol of those new physical freedoms and erotic sensations was, of course, the gas-fueled automobile, or the "horseless carriage," since it gave bourgeois Americans the sort of immediate "increase of power" and "acceleration of movement" that characterized modern geographic mobility and its corresponding sensory release.[26] This acquisition of "power and mobility," as Lewis Mumford phrased it, was in fact what made the automobile sink so deeply into the unconscious structures of modern being. By speeding up the experience of time, widening one's geographical orbit, and allowing for a rush of new physical and psychic sensations (i.e., "the elation of sheer speed"), the automobile allowed, contemporaries said, for an experience that was "felt deep in [the] bones," an experience that "fired up the blood" and sunk "into our innermost natures."[27] Poet Kenneth Fearing saw a sort of erotic allure in the "floating power" of the automobile and the pleasure of its soft "knee-action" just as others found sensuality in the visual rush of fragmented events that could be seen through the glass of a windshield, the passing pleasures and "imaginative suspense" of watching small disconnected scenes while on a drive, like glimpsing "a boy of eight . . . looking at the middle of [a] man's belly" and "a girl with one leg/over the rail of a balcony."[28] Such affective pleasures and somatic liberations associated with the automobile constituted the deep emotional structure of a fossil-fueled world in which prosthetic technologies saturated the collective bodily experience of Americans and produced a sort of affective affinity that underlay what Lemenager terms our "emotional investments in modernity." Automobility—and the sensory regimes around it—in other words, came to make up the substance of a "virtual body" that was deeply rooted in fossil fuels and that came to be grafted onto the more traditional organic body.[29]

Although it is not easy to exaggerate the influence of automobility on the modern body, it is important to remind ourselves that the middle-class body was also extended in many more quotidian ways that became part of its second nature. Fossil fuels were also naturalized through countless unspoken somatic rituals that attended the proliferation of smaller energy technologies, such as electric washing machines, water pumps, vacuum cleaners, refrigerators, blenders, fans, lawn mowers, chainsaws, and hand-held drills, each of which reshaped the experiences of labor, leisure, and eros. The example might seem incommensurate, but even a small energy technology such as the kitchen beater—or what a member of the Good Housekeeping Institute called the "mechanical arm"—extended the modern body beyond its original scope and changed the texture of life by allowing housewives and chefs to whip butter or cream without the sort of tiring repetitive arm motions previously required to make such things as bread, butter, and cake. Because the electric beater did the muscle-intensive work of mixing, beating, and whipping that women's (and sometimes men's) muscles once

The prosthetic body at work: Andreas Feininger, *Columbia Steel Company at Geneva, Utah—Bulldozer Handling Gravel for Concrete during the Construction of a New Steel Mill,* 1942. Office of War Information Photograph Collection. Courtesy of Library of Congress Prints and Photographs Division.

did, it quite literally functioned like a "strong right arm," taking some of the heft out of the daily work routine in the kitchen and redirecting women's bodies and senses in small conscious and unconscious ways. That is to say, not unlike the automobile or airplane, these other energy technologies called into play a new array of somatic experiences and embodied memories that surrounded movement, work, and pleasure in such intimate spaces as the kitchen and home. As one contemporary woman wrote, no longer were "a brawny arm, reddened hands, and an aching back" the somatic manifestations of housework.[30]

The physical and psychic releases along with the eroticism that came with extending the scope of the modern body in these ways — that is, the pleasures of becoming prosthetic — were, of course, differentiated along gender, class, and racial lines and never simply one-sided. Developing a virtual body or, to use Chase's words, operating "outside the laws of living organisms," was, in other words, not a singular experience.[31] Even the emancipatory side to modernity's ontology was always fractured socially. Yet coming into the power of a billion wild horses did restructure the texture of nearly all Americans' somatic and affective experiences as bodies shed old movements and absorbed, both consciously and unconsciously, the sensory details of those changes. Today, we might still be mapping out the multitudinous ways in which a surfeit of big energy recoded our intimate movements, emotions, and pleasures and rewrote the aesthetic and sensory history of the self, but we are at least coming a little closer to understanding what it means to be a people of prehistoric carbon down into the bone.

Big Energy and the Diminution of Workers' Bodies, or Trained Gorillas on the Line

Pleasure, catharsis, eroticism — these have long dominated the public transcript of energy, modernity, and the body. Because mineral energies supplemented and extended the organic body's own limited physical capacities, a sense of expansiveness, freedom, and pleasure became deeply rooted in modernity's ontology. We might say that fossil fuels saturated the muscle memory and aesthetics of modern life by remaking our corporeality and reconditioning our consciousness around these new liberations. Nye calls it the "tacit dynamic and perceptual knowledge" we gained from moving through the world in new ways.[32] And yet equating big energy to pleasure and emancipation belies the historical complexity of our encounter with fossil fuels, an oversimplification that has in the past enabled a sort of infantile boosterism to take shape wherein companies selling energy technologies, such as Whirlpool, could claim, as it did in the 1952 film *Mother Takes a Holiday*, that women's emancipation was never so much a question of politics as it was of electric washing machines.[33]

The fact is that modernity's ontology has always been highly strati-fied in social terms, as privileged middle-class, and frequently male, bodies like Lindbergh's expanded beyond their organic limits, while less privileged subaltern bodies diminished in scope and retreated in their encounters with modernity's energies. This second, more stress-ful embodiment of fossil fuels found its purest manifestation on the assembly line (and also in the nation's steam-powered laundries and other proliferating sites of automated production), where working bod-ies contracted rather than expanded whenever carbon was combusted. Here fossil fuels marked an important departure in the somatic life of the American worker as they helped to reset the experience of labor and coax Americans to sweat to new rhythms. To contemporaries, it was simply the "tempo of modern life," but they knew all along that the costs behind that exhilarating tempo were borne out disproportionately in the bodies of subaltern workers who experienced modernity's power in socially stratified ways.[34]

We can get a sense of the physical and psychic reductions subaltern bodies underwent when faced with modernity's energies by turning to Henry Ford's River Rouge facility, where the world witnessed the matu-ration of the sort of high-energy automated production that came to de-fine the modern work environment. With its eye-grabbing smokestacks, dynamos, boiler rooms, and complex of pipelines, industrial sites such as the River Rouge Plant were among the most important places that manufacturers experimented with the functional relationships among fossil fuels, work, and the body. As Ford put it, the two governing prin-ciples behind his own facilities were that the heaviest labor might be lifted off of "flesh and blood" and placed onto "steel and motors" and that men might subsequently be "united" with mineral-powered ma-chines.[35] Here in modernity's high cathedrals, the subaltern body was repurposed for high-energy production, a development that worked to strip it down to a sort of raw "physical animality."[36]

The bodily regimen demanded by modern steam-powered produc-tion and then later electric line production was intense. Ford, for his part, boasted that his goal was to recondition workers' bodies so that they would know deep in their bones that "the work, and the work alone, controls us," a forbidding principle when set to the logic of modernity's igneous engines. Achieving such a goal, of course, put heavy constraints on the modern body. It meant, as Ford explained, that workers had to learn to repress somatic instincts such as walking around when on the job since electric conveyer belts could carry the work to them and accomplish the same task without muscular effort and with less waste of time. Similarly, other traditional somatic liberties such as conversation, play, and conviviality were pushed off of the shop floor and cordoned off to workers' leisure time as line worker's minds and mouths were asked to clam up for the purposes of high-energy productivity. Ford boasted, in fact, that he had so completely integrated his workers into his facility's machinery that by the 1920s even what he called "substandard men," a crude term he used to refer to blind men, deaf and dumb men, epileptics, and amputees, could with the help of prosthetic energy technologies achieve the same high levels of physical productivity on the line as did the factory's "standard men."[37] Such concern with the body and with disciplining it to the metronymic pulse of modernity's energies struck at the core of traditional beliefs that humans were defined by their skills, autonomy, and creativity as makers. In the Ford factory, not only were somatic freedoms eradicated from the workplace to make way for a raw, instinctive, and recalibrated modern body, but so too was the traditional belief that work should be a form of creativity and self-realization.

Workers who were compelled to subjugate their natural and social instincts to the demands of high-energy production experienced this second embodiment of fossil fuels as a type of somatic and psychic shock. Most of those workers emphasized that they had only entered into the bargain because the employment was stable and because the pay was high (at least relatively for an industrial-era blue-collar worker).[38] But workers regularly discussed the corporeal costs of this somatic recoding of their bodies and minds. One worker explained, for instance, that laboring on the line reduced his entire body down to a series of instinctual "jerks, twists, and turns,"[39] to the sort of traumatic animality Frederick Taylor had equated to that of trained gorillas. Another explained in an interview that he felt that working on the line caused his body to shrink down until all he could hear was the sound, as he put it, of his "own heart beating." And a third more agitated worker remembered feeling trapped in his own body while working on the electric assembly line: "No talk, no walk, don't leave your job without permission, not even for a drink of water, not even to go to the toilet, not one step away from your operation."[40] Whatever the particular experience, working bodies frequently felt modernity's energies as a restriction, a constraint, a clamming up of the self—from the urethra to the brainstem—for the purposes of high-speed production. Even such advocates of corporeal discipline as Ford acknowledged that the sort of mindless, repetitive, cramped, and fast-paced discipline required of modern bodies on the electric assembly line was simply "terrifying."[41]

This contraction of workers' bodies amidst modernity's energies found its place in popular memory in Charlie Chaplin's 1936 *Modern Times*. In that film Chaplin, playing the part of an assembly line worker, enacted for his audience the corporeal costs of this new intermingling of fossil fuels, work, and the body. The key scene in the film, a well-known one, centers on a worker played by Chaplin who, when faced with the relentless pace of the electric assembly line, falls behind on the job, gets in the way of his coworkers' ability to perform their own jobs, and then is finally and literally absorbed into the factory's power machinery in a metaphoric embodiment of, or engulfing by, the machine. It is our most familiar cultural expression of the subaltern embodiment of fossil fuels, even if the somatic dimensions of the scene have not always been recognized and thus require some teasing out.

The scene begins with Chaplin working on the line for the Electro Steel Company, a company that calls us back to the primacy of electricity and steel in modern life. The humor of the scene centers on the fact that Chaplin's body, like all bodies, is idiosyncratic and resists being automated. First, Chaplin's body develops an involuntary itch in the armpit that Chaplin can't resist scratching. Scratching that itch, of course, comes at a cost. It means Chaplin has to take his hands off the job, the consequence of which is that he falls behind on the line and crashes into the worker next to him in a mad dash to catch up on his work. Next, Chaplin finds himself fending off a fly that buzzes around his head and decides to alight upon his nose, activating his nerve ends in another unwanted intervention of the body into the job. Chaplin squirms, he wiggles, and then he finally loses all bodily control as he stops working to swat the fly away in an act of involuntary somatic resistance to mechanization that this time forces the entire factory to shut down. Next, Chaplin takes a break from the line in order to fulfill another bodily need, to urinate, and to find some physical relief in a cigarette. Even here, however, the somatic costs of the line follow him to the lavatory as his body involuntarily spasms from its hours of unvaried repetitive activity. Following that short break, Chaplin then returns to the line even while his body remains resistant as ever. This time the owner of the Electro Steel Company orders that the power be turned up to the "limit." We

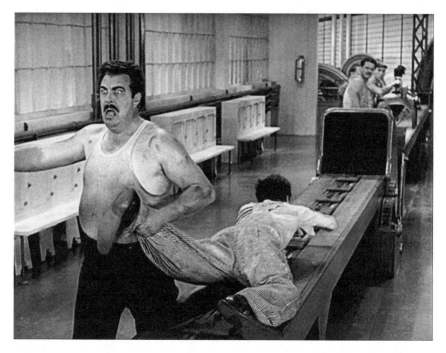

The raw and instinctive body: *Modern Times,* 1936. DVD capture. Image
Entertainment.

hear the loud hum of the dynamo, the conveyer belt picks up speed,
and we watch Chaplin unable to synchronize his body to these more relentless disembodied energies. He falls behind, he climbs onto the conveyer belt, and he finally loses his mind as he gets sucked into the gears
of the machine and becomes a literal cog in it. The cathartic laughter
produced in this famous scene comes from our bearing collective witness to the shared somatic terrors of working in such close proximity to
modernity's energies.[42]

A more sober reckoning with the somatic costs of high-energy production came in the 1908 Supreme Court decision known as *Muller v.
Oregon.* In the now-famous Brandeis brief that won that case, the de-
fendants successfully presented their argument that modern steam-
powered laundries and other such high-energy "mechanical establish-
ments" were so novel and so physically demanding that they posed a
real threat to workers' bodies. In a section titled "The Present Character

of the Business," the State of Oregon argued that the "application of power-driven machinery" in the laundry industry had so radically escalated the demands on women's bodies in the industry that they were gradually falling apart on the job. According to the state, this new high-energy work environment—with its "huge steam or gas-heated cylinders," "hot, revolving rollers," "cloud[s] of steam," and "temperature[s] of over 90°F"—not only bore no resemblance to the sort of traditional laundry work women had previously done at home but exacted from the organic body a physically exhausting routine that, according to their lawyers, endangered women's reproductive health.[43]

Distorted pelvises, impaired reproductive organs, stomach ulcers, varicose veins, crippled knees, nausea, and disoriented minds were part of the laundry list, so to speak, of the somatic costs that the state argued new steam-powered laundries were extracting from women's bodies. Although there was, to be sure, a lot of medical nonsense in the Brandeis brief, which derived from the state's use of the coarsely gendered medical knowledge of the period—one testimony claimed, for instance, that the bone structure of women's knees was so physically compromised that women should not be allowed to stand at all, let alone asked to stand for ten to twelve hours a day[44]—the argument that was in favor of regulation in this case nonetheless bore down on this distinctly modern problem of fossil fuels, work, and the body. Beneath the nonsense of gender stereotyping that motivated the verdict (which accepted the state's argument that women's bodies were weak vessels and thus in need of legal protection) was the nation's first formal recognition that American bodies were contracting, shrinking, and degenerating rather than expanding in the face of modernity's power. For a moment, that is, before the *Muller* verdict was overturned, we as a nation temporarily, and imperfectly, acknowledged in the nation's highest court of law that modernity's energies had come at the cost of a series of somatic contractions of working people's bodies. More typically, we simply complained that modernity was producing a new dehumanized race of "robots," "automatons," "iron men," and "marionettes."

Of course, to keep the matter straight, fossil fuels did not in themselves constrict workers' bodies in the ways described above. That was the cultural work of industrialists who were intent on bringing the nation's impressive mechanical horsepower to bear on maximizing productivity and profits. And, of course, workers resisted that development in their own ways. We can hear, for example, objections to the somatic and psychic sacrifices demanded by a fossil-fueled modernity in the claims of workers' groups such as the Knights of Labor (the nation's earliest organized national labor movement), which argued that physical labor should not be regarded as a type of brute power but rather as a dignified and self-realizing experience and that laboring bodies themselves should not be viewed as fodder for modernity's engines but rather as the nation's symbolic center.[45] But this is simply to say that the impact of the culture of big energy on subaltern bodies took shape, as we know, within an inequitable capitalist system that regarded biological labor, as it always had, as expendable.

Yet if the nation's laboring classes suffered both before and after the shift to prehistoric carbon, the structure of their suffering changed dramatically. The introduction of fossil fuels into the economy fueled, quite literally, the universalizing of a set of distinctly modern somatic and psychic experiences on the job that had once been limited to the now seemingly quaint watermill operations of places like Lowell, Massachusetts. Fossil fuels, that is, carried the somatic patterns of high-energy mechanized production well beyond New England's watersheds—to New York, to Chicago, to Detroit, to Pittsburgh, to Boulder, and to Atlanta—while perfecting and intensifying the subaltern body for such modernized production. The pace and swell of modernity's energies, in other words, defined in good part what it meant to be working class in the modern world.[46]

Big Energy at the Point of Its Production: The Case of the Coal Miner's Body

The embrace of modernity's energies exacted its pound of flesh, so to speak, at the point of combustion. Americans, however, also embodied fossil fuels at the point of their production—that is, along the nation's mineral frontiers, where coal, oil, and natural gas were first coaxed out of the ground. Here in the nation's and world's coal mines, stokeholes, and oil fields, other marginalized bodies knew coal, oil, and natural gas directly through the vantage point of labor. The pleasures and stresses, reliefs and sufferings, and corporeal adaptations we associate with the modern world were, in other words, always chained dialectically to these other frequently raw, maimed, and stunted subaltern bodies, which worked along the world's mineral frontier. These bodies were arguably the most modern of us all.

The body of the coal miner is the central figure in this regard. Unlike today when coal mining has been mostly mechanized and when petroleum tops the nation's energy budget, the coal miner's body was both materially and culturally central to the rise of a nation and a world built on ancient carbon. Like roughnecks and refinery men, coal miners experienced modernity's energies at the moment in which they were being pulled, cracked, and coaxed out of the nation's lithosphere. Spending much of their life cloistered during daylight hours in the ground, these subterranean bodies and their terrestrial counterparts like the coal stoker absorbed, more than anyone else did, the blunt somatic impact of modernity's dependency on fossil fuels. This embodiment of coal at the point of production, an experience that directly affected nearly 750,000 coal miners in the 1910s, or as many as one in every fifty adult male workers above the age of sixteen,[47] gave to modernity's ontology a sharp social stratification defined by cramped and deteriorating subterranean bodies juxtaposed against the corporeal exuberance and psychic emancipations of the nation's more privileged classes.

Coal miners knew fossil fuels through labor. We can get a glimpse of this other mode of embodying fossil fuels by turning to the experiences of a coal miner like Earl Brown, an African American miner who spent the good part of twenty-two years underground in U.S. Steel's coal mines outside of Birmingham, Alabama. Although Brown had known coal in relative proximity from his birth, it was not until he decided to enter the coal mines as a nineteen-year-old in 1938 that he came to know coal in its raw state. Listening in on his experiences gives us some indication of what it meant to embody coal at the point of production.[48]

Brown's first day on the job was an especially tough one for a virgin miner. He was sent deep down into an already hollowed out seam that was being squeezed for its leftover coal. On that particular day two older men were working a top shelf, trying to get a ledge of coal to fall. It was the sort of dangerous work typical of coal mining but a job usually reserved for veterans. Even a half-century later, Brown could feel the experience with a visceral immediacy:

Never will forget it. . . . [They] went to working and went to working and all of a sudden [the top shelf] fell and that's the awfullest sound I'd ever heard in my life and I just took off and went running. If they hadn't caught me, I guess I'd have run myself to death. Never will forget it. I would have quit then . . . [if the foreman hadn't put him on green work].[49]

Brown came to embody coal the hard way. The dropping of the heavy shelf, which shook the earth violently, produced a gust of wind and generated what he called the sound of "thunder, the heaviest thunder," provoked in him the somatic instinct of flight in which his heart went racing and his legs sent him running to get out of the mine until he was no longer surrounded by coal or thunder. Such a reaction was coded deep in the genes, and it could still be felt decades later. Following that incident, Brown was shifted to more appropriate "green work" and was thereafter allowed to develop the mind and the body of a coal miner more slowly. But like a frog in boiling water, he and miners like him had to learn—sooner or later—to suppress both that deep somatic instinct that sent one running when the earth convulsed violently underfoot and the full-throated anxiety that came with hearing "thunder, the heaviest thunder" overhead on a daily basis. Brown mentioned those memories more than once.[50]

As an eighty-year-old man who had survived more than two decades underground, Brown could laugh off his early reactions as that of a greenhorn, although his laughter was clearly loaded with implication. But what is clear from his recounting is that in order to spend the rest of his life working in the mines, Brown had to learn to reprogram his body to live in a world of fossil fuels. In his case that meant sublimating his deepest somatic instincts and learning to bury the always-present knowledge that coal and rock sometimes did give way and a slate roof actually did fall on someone's head. Embodying fossil fuels meant, in other words, that modernity's somatic expansions and psychic liberations were tied directly to this other set of physical contractions and repressions in the body of the miner.

Knowing coal through labor went well beyond this sort of physiological repression. It also meant the experience of breathing in coal dust on a daily basis, adapting visually to a diminished array of colors, accustoming oneself to artificial light and to short distances, cramping one's body in small wedged spaces, feeling the dankness of wet rock and inorganic matter on one's skin, and unconsciously recalibrating one's senses to an olfactory workscape wherein prehistoric carbon saturated the air with its variations of stinkdamp, blackdamp, afterdamp, and firedamp. "Such embodied knowledge," Thomas Andrews writes, amounted to a type of work-specific "consciousness that seemed innate but that was actually acquired through years of on-the-job education." A coal miner, he says, came to know coal and the coal mine along, among other things, "the

Embodying coal at the point of production: Lewis Hine, *Brown Mine, Man Digging Coal by Machinery*, 1908. Records of the National Child Labor Committee. Courtesy of Library of Congress Prints and Photographs Division.

tips of the fingers" and in the "retina of the eye."[51] This embodiment of coal at the point of production occurred at both a conscious and infra-level of perception as coal miners' bodies absorbed their decidedly unique natural environment and redisciplined themselves in minute ways to a mineral workscape for which they had not evolved.

For example, not even the air the coal miner's body took in was free of fossil fuels. Working with coal meant breathing in an atmosphere different from air on the surface. Mine air, comprised of prehistoric carbons such as methane and coal dust, moved silently, had to be refreshed with fans, was often deadly, and was only partially known, felt, and controlled. Larry Mantowich, an Illinois miner, for instance, described the quality of mine air that produced the type of scarred and inflamed lungs that characterized coal mining bodies for much of the industry's history: "We had a long time [without clean air], the dust on the cutting operation was so intense . . . so intense that respirators were impractical because about five or six breaths, why, it would block up the respirator. And your nostrils would block solid."[52] Neither biological lungs nor early respirators could filter out the type of silica and coal dust that scarred, thickened, and created lesions in otherwise healthy lungs. While the coal miner's body might have learned to adapt to most other aspects of its environment, it never developed this biological ability to respire coal dust. The dust just stuck there and blackened the lungs, slowly smothering miners from the inside out. Jack Battuello, another miner from Illinois, remembered that some men's lungs were so impaired that they "couldn't breathe anymore" nor even walk "two or three feet" without losing their breath.[53] And the musician Joe Freeman remembered his dad gradually losing his ability to breathe: "Why he'd be so tired he would just sit down by the old coal stove until bedtime. . . . He would say, 'I'm smothering, son.'"[54] This is what was meant when miners complained that their bodies were just being used up by the mines.

Black lung was perhaps the most literal embodiment of fossil fuels, but the coal miner absorbed prehistoric carbon in other ways as well. Merle Travis, the songwriter, was not far off in imagining that embodying coal was a sort of "habit that seeped into a miner's soul/until the stream of his blood is as black as the coal."[55] Every coal miner imbibed coal literally, leaving the coal mines with the taste of coal in his mouth, with coal wedged underneath his fingernails, with coal smudged on his face, hands, and clothing, with the feel and smell of coal on his mind, and with a perceptual awareness that had been shaped each day by working in an environment centered on coal. When miners talked about being black as the coal, they were referring to this deep and otherwise unarticulated embodiment.

That literal embodiment of fossil fuels at the point of production resulted, of course, in impaired lungs and other long-term physiological problems, but coal was embodied in more dramatic ways as well, such as in the broken bodies that resulted from unexpected methane explosions, fallen slate, and the various other risks that came with harvesting carbon. Every coal miner who worked in the mines knew of at least one person killed by coal, someone like George Coleman who had unintentionally entered a "gas hole," sparked a firestorm, and had his entire body burnt "beyond recognition."[56] Those burnt and maimed bodies from mine accidents and disasters (about 35,000 injuries or deaths a year in the 1910s, or about a million casualties over the course of an average miner's lifetime)[57] were part of a collective embodiment of fossil fuels that had an impact that stretched far beyond their number. Not only did they cause miners to cultivate a somatic awareness of the textures of the rock and gas that surrounded them on a daily basis so as to tune their bodies to the "cupped surfaces," hidden "cavities," fallen "steeple effects," and lurking "pockets of gas" that made up the risky workscape of the underground mine[58] but they also generated a deep and often sublimated anxiety among families and friends in coal mining towns, a collective embodied sorrow over and stress about fossil fuels that could be felt in traditional folk songs like "Dream of a Miner's Child," "West Virginia Mine Disaster, and "He's Only a Miner Killed in the Ground."

As an example, "Dream of a Miner's Child" gives us a glimpse of this collateral impact of coal on mining communities:

> A miner was leaving his home for his work,
> He heard his little child scream.
> He went to the side of the little one there,
> Oh, Daddy, I've had such a dream.
>
> Oh, Daddy, dear Daddy, please don't go away,
> For dreams have so often come true.
> Oh, Daddy, dear Daddy, please don't go away,
> I never could live without you.[59]

This deep-seated anxiety and its complementary mourning represented this other embodiment of fossil fuels at the point of their production. Here our sorrow songs remind us that coal seeped deep into the bodies and the psyche of a culture of prehistoric carbon in which small children had nightmares about their father's death, where widows mourned fallen sons and husbands, and where miners themselves trembled and prayed to God on any given day to keep "them from harm" just one more time.[60]

This embodiment of prehistoric carbon was, of course, mostly repressed in favor of modernity's more cheerful narratives of consumption, but it periodically found public expression, as it did in the 1941 Oscar-winning film *How Green Was My Valley.* That popular film about the decline of a Welsh coal-mining town centered on the crucifixion of the coal miner's body. The film's point, which was not subtle in this regard, was dramatized in a climactic scene in which a mine blowout summons the mothers, children, brothers, and sisters of this town's miners to the mouth of a collapsing coal mine to witness the death of one of their own. With plumes of smoke ushering skyward from the mine and with sirens sounding the alarm in deafening tones, this mournful gathering en masse comes to learn that several family members and friends have been trapped deep inside one of the mine's chambers from the blowout. The camera pans across the funereal faces on the crowd, and it then

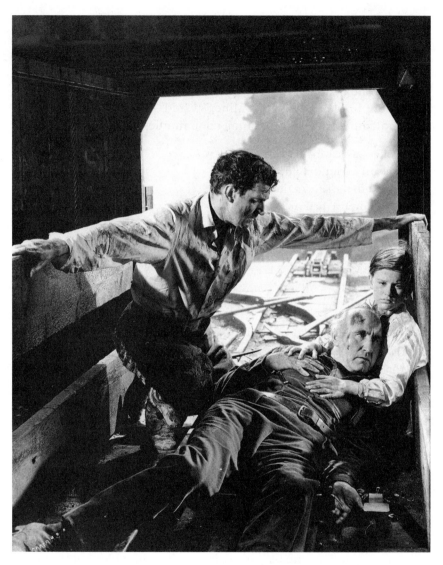

The miner's body crucified by coal: *How Green Was My Valley*, 1941. 20th Century Fox/MoviePix/Getty Images.

drops the viewer down into the mine shaft where we bear witness to the body of a coal miner encased in a ledge of fallen rock—a traumatic embodiment of coal wherein a miner's life is literally being crushed by our dependencies. In the film's final climactic image, we watch this coal miner's lifeless body rise out of the mine on a coal elevator, cradled in the arms of his son as though he is the crucified Christ, with the village priest hovering above him, arms outstretched in the form of a cross.

Here for a moment in our public discourse—in this chilling evocation of the *Pietà*—we are left to contemplate the fact that the crucifixion of the coal miner has been the ritual sacrifice we accepted for our modernity.[61]

Given the centrality of coal to the nation's development between the mid-nineteenth century (when coal surpassed wood as the nation's primary fuel source) and the mid-twentieth century (when oil started to edge out coal), it is fair to say that coal mining was, in a sense, the nation's most important profession after perhaps farming. It was certainly the profession that, as Lewis Mumford said, came to "underlie every part" of modern life, with its dark primal energy giving to modernity its light, lightness, and enlightenment.[62] For that reason, the blackened body of the coal miner might be thought of as a symbol of this other embodiment of fossil fuels at the point of production across time and region. While the experiences of a roughneck working in the early Texas oil fields were always different than those of his coal-mining counterpart, and the work of a driller at midcentury was distinctly different from that of one fifty years earlier, all along the nation's mineral frontiers—whether at home in Louisiana, West Virginia, Kentucky, Texas, or California or abroad in Mexico, Saudi Arabia, and Venezuela—this other embodiment of fossil fuels made possible modernity's emancipations. "Power," Sherwood Anderson once said, "is the coinage of the modern world." Under the terms of modernity, our power arose from these smothered bodies working in the decidedly modern, if primitive, workscapes of the nation's mineral frontiers.[63]

Big Energy and the Exosomatic Environment, or Modernity's Eroticism

Modernity's fires—from a sparkplug's ignition to the Dionysian flames of the boiler room—impressed themselves upon the modern body whenever prehistoric carbon was drilled, mined, and combusted. But they were also manifest, as architectural critic Luis Fernández-Galiano writes, in the exosomatic environment in which that body circulated, that is, in modernity's high-energy built environment of concrete and steel, smog and artificial light, and dense agglomerations of urbanized folks who lived amid the materializations of coal, oil, and natural gas. The igneous fires of modernity were expansive, in other words. They combined and congealed with nature's other matter and found their cultural articulation in a material environment that bodies came to know and absorb as their second nature.[64] If the body is indeed porous, situated, and constructed, then it cannot be disentangled from these surfaces, channels, noises, and smells of prehistoric carbon.

This embodiment of modernity's exosomatic environment was, of course, socially, temporally, and regionally differentiated. Black working-class bodies in a high-energy city such as Detroit in the 1940s, for instance, felt the materialization of modernity's energies (including its smog, its din, and its lines of steel, concrete, and brick) differently than did their white middle-class counterparts who had retreated on cemented highways to the quiet of semipastoral and yet highly networked suburbs. And yet even so modernity's bodies across the spectrum of society registered both physiologically and psychologically an external environment that was defined, in good part, by this materialization of coal, oil, electricity, and natural gas.

At the most basic and obvious level, carbon's combustion produced, for instance, physiological changes in urban bodies. Lungs developed asthma and bronchitis from air pollutants such as nitrogen dioxide, sulfur dioxide, carbon dioxide, and particulates that embedded themselves in the body's airways and internal tissues.[65] Eardrums acclimated to the high-decibel battering of squealing trolleys, excavating equipment, and muffler-less motorcycles by losing their ability to hear the lower-level decibels of their premodern rural counterparts.[66] And the urban body's orifices absorbed in untold ways a hardened environment of concrete and steel—of congealed energy—that created heat sinks, which elevated ambient temperatures around those bodies; that rose up girdered from the ground to block out sunlight, to lower bodies' access to vitamin D, and to alter the body's experiences of light and shade; and that crowded out the sights, sounds, and textures of what we colloquially refer to as nature as its products moved from the center of peoples' experiences to the margins—to the cracks in city sidewalks, to balconies of potted plants, and onto planned greenbelts. Once surrounded by the materialization of prehistoric carbon as concrete, steel, noise, and waste, modernity's bodies, in other words, came to know a different sensuality and eroticism—the basis of a different ontology—that contrasted sharply with that of their premodern counterparts who knew not so much concrete and neon light as they did hard-packed soil, the smell of tallow candles, and the scent of burning wood.

This embodiment of the exosomatic environment was an accepted fact among contemporary observers who understood that fossil fuels had done something to their environment and who did not always like what that meant. How that environment impressed itself on the modern body, of course, went well beyond purely physiological changes. It also imprinted itself on the conditioned sensuality and sensibility of Americans. Lewis Mumford, for instance, detailed some of the sensual differences between the embodiment of an "organic" world of wood, wind, and water and that of modernity's new industrial environments, which were, as he put it, an "extension of the coal mine." Here is how Mumford saw it:

> Shut off from the country by miles of paved streets, the most common sights of field and farm might be strange to them: the sight of violets, buttercups, day-lilies, the smell of mint, honeysuckle, the locust trees, the raw earth opened up by the plow, the warm hay piled up in the sun, or the fishy tang of beach and salt-marsh. Overcast by the smoke-pall, the sky itself might be shut out and the sunlight diminished; even the stars at night became dim.[67]

Clearly, while contemporaries of this transition fell back on an unreliable pastoralism to make sense of the changes they were witnessing, the sensory history of modernity's urban spaces that they recorded for us, for which coal and its smoke was a metonym, was very real indeed. From the sight of violets in the springtime to the smell of warm hay in the fall, the premodern and predominately rural body knew the world differently than did that of its urban counterpart living in the high-energy city where warm hay was limited to the feedstock of city horses. That is to say, observers like Mumford were headed in the right direction in claiming that the "erotic life" of modern folk—from their experiences of sunlight to their basest sensuality—had been adjusted in untold ways to better suit a fabricated environment comprised of congealed coal, oil, and natural gas.

Modernity's bodies absorbed, resisted, and cognitively mapped the nation's new high-energy environment in more ways than we have thus far been able to tally. The work of recent critics of technology can be

Congealed carbons and the exosomatic environment: George P. Hall and Son, *Curve at Brooklyn Terminal, New York, and Brooklyn Bridge*, 1898. Courtesy of Library of Congress Prints and Photographs Division.

useful in this regard as they have tangentially already pointed us toward this other embodiment of coal, oil, and natural gas in their discussions of bodies and machinery. As one example, we can turn to the work of a recent historian on modern music and technology, which gives us a sense of how the high-energy machine age was thoroughly embodied by one group of urbanites, in this case, jazz musicians, dancers, and their audiences during this earlier energy shift. Joel Dinerstein argues that even something as nonmaterial as modern jazz music could be traced back to an exosomatic environment saturated in the rhythms of prehistoric carbon flows. The type of "propulsive, fast, fluid, and precise" rhythms of big-band swing represented, in effect, he says, a somatic and psychic effort to wrestle with and to master the nation's "networked systems" of electric trolley ways, steam-powered train lines, mechanized conveyer belts, and sprawling highways and what he calls their unique "industrial soundscape." In the nation's jazz clubs, he tells us, modernity's besieged bodies worked to "make sense" of the noise and rhythms of this new high-energy environment and to "get with" that noise and those rhythms by capturing them and rechanneling them on the dance floor. The lindy hop, big-band swing, and tap dance represented, Dinerstein argues, a sort of inscription of machine-age values on the modern body, with their musical tropes such as the train in "Chattanooga Choo-Choo" capturing that culture's drive and flow, its power and speed, its acceleration and repetitive rhythms, and its aural associations.[68] From ragtime to free jazz, subaltern bodies, in other words, found a psychic and somatic agency in a machine-powered world that otherwise tended toward the standardization, constraint, and repression of those bodies. To put it crudely and in a way Dinerstein does not, jazz was the product of fossil fuels, and coal and steam were its most elemental inputs.

Of course, jazz was simply one embodiment of coal, oil, and natural gas; similar arguments can be made for rock and roll, futurist art, modernism, and any number of cultural forms that speak to the varied ontologies that arose in prehistoric carbon's wake. Historian of technology Cecelia Tichi, for instance, has argued that the nation's poetry, art, novels, and even children's toys each registered the impact of modernity's exosomatic environment as bodies and minds adapted to what she calls a "gear and girder" world that, as we know, rose up on an ecological infrastructure of prehistoric carbon.[69]

Tichi argues, for example, that this newly mechanized world of factories, cities, and smoke affected the deep discursive structures of

American life as people learned to absorb their new gear-and-girder world and to develop a new "machine-age consciousness" more closely aligned to that world. Although critics such as Tichi and Dinerstein are more interested in technology than in fossil fuels, their conclusions are pregnant with implications. Tichi explains, for example, that during this early energy shift, "trees, animals, and engines" (once categorically disparate objects) came suddenly to be "structural analogues" of each other as our evolving ontology rejected the older binaries of organism and mechanism that had previously structured language in the somatic age. Americans, she tells us, learned to speak their world during this first energy crisis not only by talking about gears, girders, and power technologies but also by assimilating them into a language that reflected a mechanized world.

Modernity's language and consciousness were unique to this era of fossil fuels, embracing, Tichi argues, its own artificial constructedness, indulging in the efficiency ethics of the engineer, and so freely blurring the older somatic-era distinctions between organism and mechanism that those distinctions began to quiver, wiggle, and, at times, collapse. To us moderns, she explains, coal-powered machines like trains came to have bodies with "entrails gorged with lifeblood" and gas-powered threshers acted like hyenas with "iron teeth, gnashing and threshing, devouring always, never glutted . . . snarling and slobbering." Conversely, people themselves became "machines vibrat[ing] with speed" or "just a screw or a cog in the great machine" of life, while animals' senses were "keyed to the most exquisite pitch" in this new discursive world. In other words, modernity's energies embedded themselves not only in the body's tissues but in the ontological depths of the self.[70]

"The world [is] with us." That is at least how one environmental historian has recently put the challenge to modern critics faced with making sense of our place in the world at the edge of the twenty-first century.[71] If we are to fully unpack the historical specificity of the modern self, then that means reckoning with the fact that the world that we now carry inside of us is a world of prehistoric carbon in which our suffering and eroticism, our nervousness and our bombast are at least in part produced by this absorption of fossil fuels into our muscles, memories, and instincts.

Part II of this book turns to our failure to confront that fact.

II | Submergence

3 | An Upthrust into Barbarism

For the value of coal . . . finds its true measure not in numbers or
horsepower units, but in its effect on the soul of man.
—Robert Bruère, *The Coming of Coal* (1922)

In the spring of 1922, coal was on the mind of journalist Robert Bruère, as it was on the minds of many other Americans contemplating the prospect of a summer and winter without fuel. Coal production in the United States had tripled in the past twenty years, and the expectation of 600,000 coal miners walking off the job in protest of their poor working conditions (as they eventually did) threatened to mean more than a few cold stoves and stalled engines. By the time Bruère took up the subject, coal had, in fact, insinuated itself so completely into American life that it accounted for a whopping 72 percent of the nation's energy budget and, as he explained, defined the possibilities and limits to the nation's production of "food, clothing, houses, ships, railroads, newspapers, chemicals and guns." Bruère might have been melodramatic in claiming that coal represented the material form in which "peace and ease, beauty and learning, cooperation and brotherhood" had come to the human race, but there could be little doubt that coal had changed the quality and order of life in the United States and that it was going to be a cold, hard slog to do without it.[1]

Americans had once fantasized, historian Thomas Andrews tells us, about using coal and steam to build a more just and generous modernity —"a coal-fired benevolence," as one nineteenth-century dreamer envisioned it.[2] But those fantasies had been fleeting ones. By the early twentieth century, coal and steam had already come up short on their promise. Not only had they reduced the need for hard labor among the nation's middling classes and released an unprecedented material bounty into American life but they had also triggered a host of traumatic and frequently violent events that seemed to be characteristic of the industrial age. Contemporaries thus tended to see in coal not modernity's promise but a scarred and dyspeptic industrial world that had, as they said, fueled "old hatreds" with "new intensities," habituated Americans to the "wreckage and debris of the industrial city," and generated a "lowered morale" among a people chained to the assembly line and the pulse of coal's power. If the true measure of coal was to be found in its effect on the soul, as a contemporary like Bruère believed it should, then the moral reckoning with fossil fuels was a tragic one.[3]

What is most telling about Americans' relationship to coal in these years was that the nation's material dependencies on coal and its cultural valuations of coal ran in opposite directions. Despite the fact that coal served as the hard material fulcrum of the nation's modernization for more than half of a century, coal as a cultural object came to symbolize not so much modernity's very real material rewards (albeit rewards differentially shared) but rather the myriad material and psychic traumas that came with transitioning into a new industrial world. By the early twentieth century, few Americans could stretch their minds to associate coal with peace and ease, beauty and learning, cooperation and brotherhood. Rather coal spoke to a wounded and repressed part of themselves that had been buried in the unconscious, cordoned off socially and spatially in the subaltern and subterranean locations of the coal mine and stokehole, and sequestered mostly out of sight, symbolically and psychically, in a cheerful modernist narrative of progress, consumption, and growth.

To get an initial glimpse of that repressed — and unnamed — part of ourselves, we might reflect on a painting like Harry Sternberg's 1938 *Coal Miner and Family*, a Depression-era work from a larger series on coal and steel. That painting, which is accusatory, confronts the modern viewer with the cramped and dirty body of a coal miner crawling literally underfoot, crouched in the basement of the world in a literal embodiment of his subalternity, while above ground his emaciated wife and children — these depressing by-products of the bleak and ugly business of fueling our world — sit dejectedly on the surface to unsettle any simplistic notions we might be carrying around about the relationship between coal and modern progress. In a confrontational work like Sternberg's, the harsh realities of human sickliness, psychic pain, and desecrated environments become the grotesque embodiment of our energy dependencies.

Or we might turn the clock back a few years earlier to a painting such as Charles Burchfield's 1918 *White Violets and Abandoned Coal Mine* for a second representation of this morbid relationship that we have to coal. A static and depressing watercolor of an abandoned coal mine, one of the waste products of modernity, Burchfield's painting reduces the modern world down to a mute hole in the earth. The symbolic emptiness—the impenetrable blackness that we find at the center of this depleted coal mine—speaks, we might say, to the collective feeling of loss or even uncharted trauma that has always been part and parcel of the making of the modern world. Here we sense in this symbolic muteness and in what one observer calls the dying "face-like blossoms" that surround this abandoned mine, that "God, or Satan, is lurking" somewhere in judgment of—or with perverse pleasure for—this other world fossil fuels built.[4]

Coal and the Industrial Unconscious

For more than half of a century, coal carried a heavy material and symbolic burden in American life. In terms of its material significance, coal almost single-handedly provided the energy base of the United States for more than half of a century—from 1885, when coal displaced wood (and hydropower) as the nation's predominant fuel source up until 1951, when coal was, in turn, displaced by petroleum. During that critical period in the nation's history, a period that coincided not accidentally with the United States' emergence as a global power, coal constituted nearly three-quarters of the nation's entire energy budget and fueled nearly the entire economy for a few of those decades. By way of contrast, our dependency on oil has never reached higher than 50 percent.[5]

Coal was, moreover, as we saw in Chapter 1, a game changer, lying at the mucky material bottom of nearly all of the revolutionary changes that we associate with the nation's modernity, from the radical transformation in the nature and meaning of work that attended full-scale mechanization to the catapulting of the world's population and gross domestic product (GDP) to what had been under the terms of a somatic regime literally impossible heights. Put in brutish terms, coal and the Dionysian spaces associated with coal—such as the coal mine, the furnace room, and the assembly line—catapulted the nation into the modern era.

Yet for all that coal did to underwrite the sweeping material changes that characterized the modernization of the United States, coal was associated most often with the trauma of living in this brave new world of

Harry Sternberg, *Coal Miner and Family,* 1938. Collection of Sandra and
Bram Dijkstra.

cities, smoke, and power.[6] Although the tight metonymic intimacy between coal and modernity's psychic, social, economic, and aesthetic dislocations can be traced back at least as far as the publication of Charles
Dickens's *Hard Times* in 1854, those connections congealed in the
United States sometime in the late nineteenth and first half of the twentieth centuries into a popular counter narrative to modern progress that
spoke of the nation's (and world's) break into modernity as a distressing
and even barbaric thing—a "half-barbarism," as Nietzsche phrased it
in regard to the European experience. Coal as a cultural object, that is,

evoked the industrial unconscious of modern life, serving as a reminder to Americans that beneath the surface of the nation's veneer of steel, glass, and light and beneath the exhilarating speed at which its factories, ships, and railroads churned, there stood this other implacable, bare-faced, and hidden material world with its furnace stokers buried in the depths of steamships, its miners sequestered in mountain hollows, its dirty ashes hurried out the back door, and its smoke and slag removed beyond the sight and smell of the nation's middling classes. Nothing, that is, was darker, dirtier, or more decidedly modern than coal, and nothing evoked so explicitly the traumas of modern life as did this gritty black rock and the trace of its lingering smoke.

Charles Burchfield, *White Violets and Abandoned Coal Mine,* 1918.
Albert-Knox Art Gallery, Buffalo, New York.

To be sure, coal and its residue, smoke and slag, did not stand alone
linguistically but served rather as tropes in a wider social literature of the
period that centered on the ills of industrialization and that spanned the
writings of realists, naturalists, muckrakers, and sometimes modernists.

Coal's role in that literature—from the original text in that tradition, typically considered to be Rebecca Harding Davis's *Life in the Iron
Mills* (1890), to midcentury texts such as John Ford's 1941 Best Picture,
How Green Was My Valley—was to serve as metaphor and metonym for a
sublimated trauma in modern life that had been relegated mostly to the
fringes of the national imaginary. Coal called Americans back, that is, to
a disquieting and mostly unacknowledged schizophrenia in the national
self that included a stark social bifurcation between producers and consumers (and the different spaces they inhabited) and an equally stark
psychic division between the public transcript of the nation's modernity
and its raw, damaged, and antiaesthetic underside that was to be found

in the far reaches of the coal mine and in the brute materiality of the object itself. As Davis put it, billows of black smoke, the trace of coal's combustion, were the "idiosyncrasy" of this age—the backdrop to the rest of the nation's story.[7]

Huge, Uncontrolled, and Diabolically Strong

An essay such as Sherwood Anderson's "My Fire Burns"—written at the peak of coal's influence in 1922—captures both the material traumas associated with coal and the birth of the modern world as well as the middle class's persistent fear that modernity's sometimes unruly energies were eroding the material foundation of American individualism. As Anderson put it, "The America of today is not the America of a very few years ago."[8]

The sentiments expressed in Anderson's essay exemplify this broader moral indictment of coal, which spanned the writings and work of other prominent figures, including Lewis Hine, whose photography of children in the West Virginia mines tied coal to the most wretched of class traumas in modern life; Gifford Pinchot, who saw in coal a symbol of both industrialization's profligate waste and the undisciplined rise of dirty and unhealthy cities and factories; Lewis Mumford, who claimed that "mankind" had behaved like a "drunken heir on a spree" after coming into its new energy inheritance; and many lesser intellectuals who looked to coal as a symbol of both modern alienation and the collective regression to "primitive instinct" in a high-energy age characterized by "pugilistic" encounters, "boisterous" wars, "jazz dancing," and "he-man" movies.[9]

Taking its cue from this culture at large, "My Fire Burns" foisted onto coal a heavy symbolic burden. Anderson held coal accountable for the full laundry list of modernity's traumas. He explained that whereas prior to coal Americans had lived in a natural, less discordant world of trees, grass, and living animals, they had woken up to find themselves crowded into congested cities and surrounded by concrete and dirty industrial alleys; that whereas, before the coming of coal, the nation had expressed itself in an organic lyricism and beauty characteristic of the pacing of a preindustrial world rooted in human bodies and handicrafts, coal had hurled them headlong into a quantitative age of machines and mass production that was less in keeping with humanistic values; and that whereas, before the coming of coal, the nation had been populated by sturdy autonomous individuals who thought and acted with a degree of independence, coal had repeopled it with "slaves" to an "unbridled" material force that was driving them forward without plan or reason. "Show me," Anderson wrote, "wherein all this tickling of the lining of the Earth's stomach until he becomes ill and spews forth these millions of tons of coal . . . has brought anything at all of light or beauty or meaning [into the world]. . . . This breathing of smoke and black dust, this quick throwing up and tearing down of cities, this thing we so grandiloquently call America's Industrial Progress . . . has no meaning at all."[10]

In sketching this grim portrait of what had been lost in the transition to the modern age, Anderson invoked—as his peers often did—a series of archetypal preindustrial memories that called up a more "natural" and less hectic sensory world. He described, for instance, listening to "recited poetry" on a neighbor's front porch in a small town, looking up at the "blue-black night sky" in the country and seeing stars rather than lights, listening for the "sharp sound of hoof beats on a distant road," and simply enjoying the rhythm of a world where people still recognized each other by the cadence of their "gait" and by the "peculiar swing" of their arms. To represent this other prelapsarian world before the coming of coal, Anderson settled on the image of a "ragged Negro" boy on a "bony horse" delivering coal by cart and singing out in the streets as he went by. That image of a formerly innocent world when song and labor went hand in hand and when coal was simply another word for heat (however troubling its racialization) denoted to Anderson a time when labor was still done by people rather than machines and when the world moved, like a bony horse, to a slower, more deliberate rhythm.[11]

In essays such as Anderson's, coal bore the full weight of these psychic traumas that he and his peers equated with industrialization, but coal also symbolized the structural victimization many middle-class Americans felt to be characteristic of the industrial world in which they lived. As for Anderson, he used the term *slave* to indicate that feeling of impotence, but he also drew on the familiar metaphor of coal as "King Coal"—something of a superhuman figure bigger than humanity itself—to make the same point. That well-worn metaphor of the period, which carried intimations of an earlier antebellum discourse of King Cotton, expressed a popular belief that the status of the liberal individual was diminishing in the wake of this new colonizing power. The way Anderson put it was that King Coal—having been awakened from his slumber in the hills—had gotten beyond humanity's control, had broken its "bridles" and had gone about tromping willfully and "Hun-like" across the nation while "breathing his black breath over greater and greater stretches of green country." This conquest by coal had, in other words, not empowered men and women as its boosters claimed but had left them with a haunting feeling that their lives were "meaning-less" and "futile" against the backdrop of this "huge, uncontrolled and diabolically strong force."[12]

As its rhetoric suggests, this moral indictment of coal was steeped in a nostalgia characteristic of Anderson's class, but the sentiments he expressed were widespread ones as the popularity of the 1941 Oscar-winning film *How Green Was My Valley* suggests. Taken seriously, those sentiments amounted to a heartfelt, if inchoate, critique of the nation's myth of progress that found focus in this symbolic censuring of coal. Al-though Anderson's homegrown critique of industrialization might have lacked the intellectual coherence of contemporaneous Marxist writings, it spoke powerfully and directly to a mainstream interpretation of the trauma of modern life that was filtered through Jeffersonian fantasies of freedom and autonomy and a pastoral unconscious that seemed threat-ened by the sharp changes that had come with the shift to coal. To Anderson's generation, a republic built on the principle of the autono-mous individual and born in an age of foot traffic, handiwork, and local autonomy seemed at risk of being bowled over by the vast energies of industrialization. This new world order, as members of his generation saw it, was far less suitable to republican virtue and far more likely to pro-duce a "new kind of men" whose bones and sinews knew only concrete, assembly lines, and "unhuman" cities and who were being naturalized to their fate as "slaves" to a material force they could not fully control or comprehend.[13]

A Kind of Civilization within a Civilization

Not all of the period's indictments of the Age of Coal were packaged in such sentimentality. This same story of declension found sobering expression in the writings of muckrakers who visited the coal regions themselves. In their writings, the nation's coal towns and the conditions of the working class in these towns spoke not so much to the middle-class perception of trauma in the industrial age but rather to a hidden schizophrenia in modern life that was to be found in the stark opposition between these morbid working-class geographies and the more celebrated bourgeois spaces of the modern world.[14] These coal towns, with their damaged peoples and their distressed environments, appeared in the muckraking literature of the time as evidence of an ugly underside to the nation's modernity that had been removed from the sight of the nation's bourgeoisie and pushed to the far reaches of its collective consciousness. Progressive journalist Winthrop Lane referred to these coal towns as a "kind of civilization within a civilization" in the United States—an "alien order embedded in democracy."[15]

Lane's writings on the coal mines of West Virginia, published in the *New Republic,* the *New York Times,* the *Survey Graphic,* and the *Nation,* spoke to this barbaric facet of the national self. They reminded middle-class Americans of the systemic class traumas of modernity that had otherwise been buried deep in a liberal narrative of progress that downplayed class fractures and celebrated the emancipatory qualities of the nation's modernization.

Lane cast the story of West Virginia as a localized and working-class version of the story of national declension that we saw described in the previous fictional example. He explained that with the arrival of the coal industry in West Virginia after the Civil War (at the moment the nation was converting its energy infrastructure from wood and water to this new fuel source), the quality of life in the state deteriorated rapidly. Almost overnight, he said, this quiet mountain state went from being a "sleepy state, isolated, without railroads [and] with few inhabitants"—the type of place where "life on the whole was simple" and the "earth reposed peacefully"—to a national sacrifice zone. West Virginia might have once been home to a type of durable, mountain folk who took pride in their fierce independence and who played such an important role in the nation's mythology in its precoal years, but with the coming of coal, the state no longer produced such independent peoples. Rather the new West Virginians were a new type of subaltern class of men and women whose lives were mired in what Lane called a "semi-feudal" social order. "The blackness of coal is over everything" in these parts, he wrote, and he meant that both literally and figuratively.[16]

Lane did not have to stretch the truth to make his point. The coal regions were perhaps the most striking evidence in the late nineteenth and early twentieth centuries of a structural morbidity in American society that stunted the lives of the nation's working poor. These were places where the discourse of modernity and the nation's myth of prog-ress rang hollow. In the Guyandotte River region where Lane visited, for instance, the coming of coal had completely restructured the state's economy around this single extractive resource and concentrated eco-nomic power in the hands of a few corporations that owned the rights to that resource. In the case at hand, that meant two distant corpora-tions—U.S. Steel and a subsidiary of the Pennsylvania Railroad, the Norfolk & Western Railway Company. This acute concentration of power in the state—characteristic of nearly all of the nation's coal fields—translated, Lane explained, to almost complete socioeconomic control over a quarter of the state's population and to indirect politi-cal control over local and state governments besieged by the coal in-dustry's deep pockets and active lobbyists.[17] As one of Lane's peers put it, West Virginia's once-fierce Hatfields and McCoys had been reduced to "company-owned Americans," people whose lives were determined from cradle to grave by the coal industry and its stockholders.[18]

Lane and his peers were quite voluble on this point. They saw the nation's coal towns as evidence of a barbarism in American life that had been ignored by the urban middle class, even as it drew its power from these damaged regions. The central trope in this discourse was the coal town itself, which was depicted as lacking all of those things that made modern society dynamic and fluid. "No one owns his own house" in the coal towns, Lane wrote.

> No one runs a store, operates a garage or sells groceries or haberdashery to his fellow townsmen. No one amuses themself [*sic*] in a movie theater. There is no Main Street of small independent businesses, owned by different peo-ple, and making up that mosaic of commercial life that is typical of villages everywhere. There is little if any participation in common, group activities. No body of elected councilmen ever passes on repairs for roads. [N]o group of people ever gets together and decides that the old schoolhouse is too ramshackle for the children or that the old church needs repairing. No fam-ily physician builds up a successful practice by competing with other physi-cians. No lawyer settles disputes over property rights among his neighbors.[19]

The brute fact of life in these depressed regions, Lane said, was that hard-working Americans were stuck in a chronic state of dependency that was decidedly out of synch with the public narrative of the na-tion's modernity. Deprived of control over their destinies and stuck in conditions akin to feudalism, these otherwise average Americans, who numbered nearly three-quarters of a million men in the nation's work-force, stood as a thorny challenge to the nation's dominant narrative of progress, a narrative that centered on the catalytic power of hard work and individual responsibility and that had always held out the promise of social mobility.

No single writer—not even a forceful one like Lane—ever managed to capture in words the symbolism of these marginalized regions so effectively as did documentary photographer Lewis Hine, whose arresting portraits of coal miners in these regions gave visual expression to the class traumas Lane could only attempt to describe. A powerful photograph like Hine's "The New West Virginians," which centered on a portrait of motley miners—begrimed in coal dust, deep in the bowels of the earth, emotionless, excepting a vague look of stupor, or repressed aggression—pronounced in no uncertain terms that coal had carried the United States a long way from its founding fathers, yeomen, and frontiersmen.[20] In such a context, it is easy to imagine that contemporaries saw the coal regions to be evidence of some sort of preindustrial barbarism that had persisted into the modern age, but, to be perfectly clear, coal represented to them the sign of a decidedly *modern* atavism that had been coaxed into life by the energies of industrialization. In the chilling words of one journalist, these coal towns existed precisely to provide the "furtive . . . blood transfusion" needed to keep modernity's other more privileged half alive.[21] That uncomfortable truth about our relationship to coal can be felt in Anderson's claim that visiting these regions had left him with a "kind of raw tender place that ha[d] never quite healed."[22]

A Dark Bourgeois Sublime

To this point, we have seen that coal as a signifier was cathected with both the psychic and material traumas of entering the industrial age and that, as such, it circulated as a powerful trope in a counter narrative to the nation's modernity that highlighted the human costs of modernization and that brought attention to the class omissions and symbolic elisions within the national imaginary. To some extent, it might seem that this public discourse of coal was not all that different from the more generic story of decline told earlier in the century about the water-powered mill towns and the first stirrings of the industrial order in New England. The mill towns were, after all, also tied up in class and psychic traumas even if today a churning water mill appears quaint by comparison.[23]

But the story of coal was different in that coal spoke directly to a part of the modern self that had been repressed socially, psychically, and symbolically in the nation's self-portrait and that appeared as a sort of traumatic knowledge about itself when it did break through to the surface of consciousness. The next few examples, which speak to this notion of coal as a type of traumatic knowledge, illustrate, among other things, the difficulty the nation's middle class has long had in articulating, and confronting, these dirty material origins of its modern self.

Stephen Crane's "In the Depths of a Coal Mine," written for *McClure's* magazine in 1894, sheds light on this complex and tortured relationship the nation's middle class had to coal in these years. This essay, which described the author's descent into a coal mine near Scranton, Pennsylvania, framed the coal mine (and its subaltern class of workers) as a dark and terrifying underside to middle-class life on the surface. Coal called up quite explicitly in this essay a dark facet of the bourgeois unconscious that could only be glimpsed in the previous examples. Moreover, the editing of the essay itself—which underwent considerable revision prior to publication in *McClure's*—provides evidence of an additional layer of psychic and symbolic unease with the middle-class encounter with coal that was otherwise erased in the published version of this text and elided, more generally, in the public transcript of the nation's modernity.

Crane portrayed the coal mines of eastern Pennsylvania as a dark Gothic abyss in the modern world, and he cast his own descent into that abyss as a figurative descent into a netherworld the likes of which would have been familiar to Homer, Virgil, or Dante. The language Crane employed in this regard was explicit. As he approached the coal regions, the landscape, he said, became "dark," "somber," and drained of life—the vegetation "miserable, half-strangled" and the earth productive of only a "few unhappy trees." Towering over these regions were huge and sinister coal breakers that looked like "enormous preying monsters, eating of the sunshine, the grass, the green leaves" and churning out "tremendous heaps of waste" while filling the air with "infernal dins." In view of the mine itself, Crane said, two ghostlike "apparitions" appeared "from the center of the earth" and greeted him as he readied himself to enter the earth's "sunless depths." And once inside the mine, Crane claimed that he found himself surrounded by a metaphorical hell filled with the "snarling" of wolves, the "grinning skulls" of haggard miners, the "bleached bones" of black "imps," the "strangely satanic" and "ghoul-like" movements of "mysterious figures," and the "weirdly shrouded" bodies of a subterranean race of men who spoke in broken vernacular and who went about their business with an air of "villainy." Moreover, and finally, Crane said that he felt an acute sense of disorientation in the mine as he plunged into its darkness. He worried over the "threat of eternalness," and he reached out instinctively for "some coherent thing, some intelligible spot" in this "swirling dark chaos."[24]

It does not take much to see that Crane portrayed his plunge into the coal mine as a descent into an archetypal underworld populated by a familiar set of fears and fantasies concerning chaos, death, and eternity. The language he used to describe the otherness of this working-class space was the standardized imagery of hell, the grotesque, and the Gothic. Likewise, Crane cast the narrative aspect of his experience as a transgressive act in which his encounter with a socially marginal facet of his world (in this case the coal mine and its coal workers) called up images of the irrational and the repressed.[25] That is, Crane imagined that the act of descending into the coal mine (this very real working-class space) was a type of transgressive—and yet redemptive—act, not unlike Aeneas's and Oedipus's descents into the underworld, in which plunging into the abyss yielded up a measure of self-knowledge that the author would bring back to the "cool fresh air" and the sun's "glory of gold" of the surface.[26]

The coal mine called up, in other words, more than the conventional images of hell and the Gothic; it also called up a dark, dreamlike, and transgressive state of being where light, reason, time, and space were no longer the operative terms and where, as Crane put it, the "mind becomes a confusion."[27] In this respect, the author's literal plunge downward into this dark working-class geography functioned also as a figurative plunge into the unconsciousness of the bourgeois self. By framing the coal mine and its people within this familiar language of the unconscious and the low, Crane's essay thus expressed in psychically distorted form the underlying fascination and repulsion that the nation's bourgeoisie had both to this dark material object that was so central to their lives and to the damaged underclass that was associated with it. Coal was, in other words, more than simply congealed energy to them.

In depicting the coal mine as a dark bourgeois sublime, Crane relied on a familiar romantic typology that permitted a degree of psychic distancing from the actual traumas he encountered in the mines. But Crane's experience was more complicated than that. In visiting the coal mine, Crane was also visibly shaken up by his encounter in a way that was not captured in the published version of the piece. According to both the photographer who accompanied Crane on his assignment, Corwin Linson, and the author's later biographers, Crane left his assignment in the mines with a "simmering outrage" and a sense of "indignation" over the conditions he saw there.[28] Originally, Crane had vented that anger onto the pages of the article he had written for the magazine, decrying the fact that "miners were being exploited by coal-burning consumers" and railing against the predatory class of coal brokers — "those men who make a neat living by fiddling with the market" — who were profiting off of the suffering in these unhappy regions. Crane even indulged in that original version, at least momentarily, in a destructive fantasy in which the coal mine crushed a group of such coal brokers as a penalty for their sins.[29] And yet, that raw angry indictment, which ran for four paragraphs in the original draft, was left out of the published version — the latter of which retained only a mild trace of that anger in its concluding sentence, which referred to coal as a "black emblem of greed."[30]

The censorship or editing of this piece[31] for publication in *McClure's* radically changed the nature of its content, muting the uncomfortable class traumas of the coal regions and opting for a romantic construction of the coal mine that was more in line with a bourgeois aesthetic and sensibility. The final version of "In the Depths of a Coal Mine" was consequently no longer really about the coal industry or coal workers so much as it was about the psychic composition of the bourgeois self and its troubled relationship to this business of coal. The published version of the piece was, that is, the familiar story of the urban flaneur venturing into the dark reaches of a working-class geography in order to fulfill some aching desire for danger and thrills.

As for the reader who went into this piece looking for a less aestheticized portrait of the coal mines, he or she was treated to a long, literal, and earnest digression on the suffering of a mule named "China" whom Crane said was forced to live out its bleak and tragic life in the mines without recourse to the sun's rays or to fresh green fields.[32] That transference of the trauma of the coal mine from a human subject to an animal one in this piece made the horrors of the mine much easier to process for the reader, and it ultimately left the coal mine safely sequestered from life and politics on the surface. In a sense, Crane's published essay left the middle-class consumer, who was directly implicated in the original version, free to find the lessons of the coal mine somewhere other than in the inequities of industrial-age capitalism or in the dark side of a new consumer society that traced its origins back to this place. The next two examples confront this world of coal more directly than did the published version of Crane's piece.

In 1909, a decade and a half after Crane wrote "In the Depths of a Coal Mine," a freelance journalist named William Beymer took a similar descent into the coal regions to write about the experience for another major national magazine. Beymer's essay, published in *Harper's* under the title "In the Earth Beneath," resembled Crane's account in that its author framed his journey into the coal mines as a similarly figurative descent into a working-class "underworld" filled with "fancy . . . mystery . . . wonder—and fear."[33] But Beymer's text added to the bourgeois encounter with coal a rare degree of self-consciousness that reflected back on its tortured psychology.

Writing much later than Crane, after a literature on coal was already established, Beymer went into his assignment in the mines with a heavy set of expectations. Beymer said that he recalled reading other accounts of writers descending into the coal mines and that he expected his own experience to answer to certain literary conventions laid out for him. He told the reader that he anticipated, for instance, peering into the dark mouth of the mine, turning his shoulder from the surface world he knew and loved, casting one last dramatic gaze at the sun and perhaps a final flower, and then leaving civilization behind to plunge into this dark netherworld, from which he might never return. Beymer's actual experience, however, disrupted that romantic narrative. In the course of his time in the mine, Beymer found that the troubling life of coal miners and the very real dangers of the mine were not well suited to the genre of romance. Rather than fancy, mystery, wonder, and fear, the mine produced in him, he explained, a sense of unease and alienation.[34]

Beymer was disappointed, at least initially, with his descent into the coal mine. The anteroom of the mine, he said, where the elevator (after a shockingly quick descent) first dropped him, turned out to have none of the mystery or wonder he expected. It was all too familiar. Lit by electric lights rather than some mysterious primitive method, open-aired "like a big-damp warehouse" rather than a cramped crevice in the earth, and replete with a chalkboard that looked too much like it had come out of a "stockbroker's" office, the coal mine turned out to be, he said, "monotonous," "routine," and "prosaic" rather than thrilling. In this sense, Beymer experienced the coal mine as miners themselves did, as quotidian and banal.[35] No one who had actually worked in a mine could have confused it with the world of fantasy that originally framed Beymer's expectations. But as Beymer and his guide wended their way deeper into the mine shaft through a dizzying array of tunnels and as the mine turned darker, damper, and more dangerous, the novelty of the experience released Beymer from his initial feeling of having been cheated.

To describe the more romantic part of his journey, Beymer mimicked the imagery that we have seen in Crane's essay. Light, the purest sign of civilization and reason, receded as Beymer descended into the "never-changing dark" of the coal mine. Human bodies, once straight and erect on the outside, became grotesque as they stooped to accommodate low ceilings and as they were distorted by shadow. Men's speech and reason, the signs of culture, receded as "animal instinct" and the sound of "hissing" came to fill the tunnels in which they moved. And, as with Crane's experience, the mine called up for Beymer images of saints, sepulchers, and God as he neared what he called the last "outpost" of human civilization, where a few lonely and stooped figures were chipping away at the ends of the earth.[36] Part of Beymer's descent into the coal mine was, in other words, like Crane's, a figurative descent into an irrational space within the modern bourgeois self.

But inside the mine, Beymer also felt an acute sense of alienation from himself. He said he "felt as two" inside the mine — in part the "man who had lived in the world of every day with the commonplace sights and sounds of life" and in part a new and subterranean "Stranger" to that self. The mine forced him, he said, to look at himself and at the coal mine through a new pair of "strange staring" eyes. It turned out that the very real dangers and traumas Beymer saw in the coal mine caused him to feel a sort of double-consciousness while in it. From this other perspective, the language and the expectations he had brought to the occasion seemed to be wholly inadequate, if not a little embarrassing, and his extraneous presence as a bourgeois visitor in this working-class geography seemed less a romantic journey of the mind than an act of intrusion into someone else's work space. From this new perspective, Beymer claimed that the language of "fancy and wonder," along with his naive notions about descending into the mine, made him feel like a "mischievous boy" getting in the way of the "man-work" of adults.

Beymer might have come to the mine hoping to stare a sort of symbolic "death" in the face for his readers' edification, but he came to realize in the course of that encounter that concrete death and real human suffering occurred in this place. In short, Beymer entertained the notion of writing a romance of the coal mine but found that the conventions came up short. He opted instead to conclude his article with an earnest meditation on the risks and dangers faced by actual miners and on the strength and masculine dignity with which those miners daily faced the challenges posed to them.[37]

Writings on the coal mines such as Beymer's offer us a glimpse of the modern bourgeoisie's uncomfortable relationship to coal and of its unconscious suppression of the intimacy of that relationship. The estrangement the author felt in the mine attests to the strong ambivalence that these earlier generations had to this dark resource that was so central to their lives. The split self of which he spoke, and which he divided between a middle-class self on the surface and a working-class self in the mine, referenced, that is, a dark and disturbing—social, psychic, and symbolic—bifurcation in the modern world that was characterized by an unsettling, and mostly unacknowledged, division between producers and consumers, the underclass and the privileged, and the different spaces that these folks inhabited. Although Beymer might have tried to "link" these two estranged sides of the modern self together in his essay,[38] what he really offered the reader was a fleeting, if honest, impression of the emotional discomfort that a relatively privileged journalist might have had in confronting a repressed facet of his own world. The sanity of the modern self has, after all, depended on systematically forgetting this dirty material underworld, where the messy origins of that self lay buried.

This Unknown, Abysmal Brutality

A surprisingly explicit example of this repressed trauma can also be seen in the 1921 play *The Hairy Ape*, written by the Pulitzer Prize–winning playwright Eugene O'Neill that centers on an imagined confrontation in the boiler room between these two sides of the modern self. In this well-known text, O'Neill forces his audience to confront the social and psychic schizophrenia characteristic of the industrial age by imagining a moment in which this raw, damaged, material underworld of coal bursts violently through the surface of modern life to effect a rupture in its social and symbolic order. Coal does not stay put in this provocative text but instead rises to the surface of modern life in the form of a type of traumatic knowledge about itself. Even more directly than Beymer's essay, O'Neill's text, written at the peak of coal's influence in the United States, confronts head-on the class and psychic traumas buried in the modern landscape of power.

The Hairy Ape is a familiar text, although one not typically classed together with the literature on coal. The connection to coal in this play is, however, deep. The play gestated in 1921–1922 in the midst of a nationwide coal strike that sent 600,000 coal miners to the picket lines, and the play took as its setting and subject both a physical location and a class of workers directly connected to coal. The first act of the play, which takes place in the boiler room of a modern steamship, locates the action of the drama within one of the key sites of the modern world where the world's coal was once processed into the power and speed that made up the rhythm of industrial life, and the plot of the play centers on the tragic decline of one of the coal stokers who emerges from that space. As in the previous examples, coal as a signifier carries a heavy symbolic burden in this piece, serving as the sign of a repressed underside to modern life in the United States and dramatizing what many contemporaries saw to be a disturbing atavism in the modern industrial self.

The main storyline of *The Hairy Ape* concerns the protagonist Yank, a rough, muscular, and Neanderthal-like man who works in the boiler room of a transatlantic steamship. Yank serves in this text as coal's representative. His job as a coal stoker positions him in close proximity to coal, and the text figures him as something like coal's human incarnation. Yank claims to "eat up de coal dust" and to "drink in de heat" in the opening scenes of the play, and he refers to himself thereafter as coal's incarnation in speed and steel: "Twenty-five knots an hour, dat's me!"[39] This connection between Yank and coal is developed most fully in the protagonist's opening monologue:

> De engines and de coal and de smoke. . . . Dat's fresh air for me! Dat's food for me! I'm new, get me? . . . I eat it up! I git fat on it! It's me makes it hot! It's me makes it roar! It's me makes it move! Sure, on'y for me everyting stops. It all goes dead, git me? . . . I'm de end! I'm de start! I start somep'n and de woild moves! . . . I'm smoke and express trains and steamers and factory whistles. . . . And I'm what makes iron into steel![40]

Yank serves, that is, as a personification of the heat, the light, the motion, and the speed of the modern world, or, we might say, a metaphorical embodiment of the "liberated energy" that O'Neill sees lying at the bottom of modern civilization. In his own words, he is what makes the "woild move."[41]

As a coal stoker responsible for providing the power of the modern world, Yank also represents an integral, if repressed, facet of the modern American self—but a part of that self that has degenerated into a grotesque form. His name, Yank, and his explicitly white identity give him secure claims to a privileged genealogy within the American self that descends from the New England patriot, while, at the same time, his simian qualities (he is described as "stooping," "hairy-chested," and "Neanderthal") indicate that he, and this new breed of coal stokers of which he is the purest expression, "their most highly developed individual," represents an unsettling devolution of that self—an upthrust into barbarism, so to speak.[42] This atavism of the modern American self that we see in Yank is represented, moreover, as a distinct product of the nation's modernity and its emancipated energies rather than some holdover from the past. This is clear from the setting in which he operates, an explicitly modernist space—a cage of "white steel" with nothing "naturalistic" about it—and from the temporal framing of the play, which, in at least one version, sets Yank firmly within the "modern."[43] Yank is, in short, the manifestation of a traumatic degeneration of the national self that is tied to the nation's adoption of coal.

The central dynamic of O'Neill's play turns on the fact that Yank and the stokehole do not exist in isolation but rather represent half of a decidedly bifurcated world that is split socially and spatially between different classes of consumers and producers and split symbolically between the public facade of modern culture and the raw materiality that supports that culture. The coal stokers and the furnace room represent, in this respect, the dirty material half of that world—the primal and repressed Dionysian energies from which the other more civilized half draws its life force.

O'Neill makes that association clear both by the physical location of the stokehole on stage and by the imagery that he uses to describe it. As one might expect, the coal stokers and the coal ovens are positioned physically below the public promenade of the ship where the bourgeois passengers lounge in the open air (a positioning that indicates their physical and spatial subordination in modern life); and the

Confronting the boiler room: Eugene O'Neill, *The Hairy Ape*, 1922. Abbe Studio, New York. Yale Collection of American Literature, Beinecke Rare Book and Manuscript Library.

stokehole is cast in the familiar imagery of the low and the repressed that we have seen in previous images of the coal mines: its coal stokers are suspended in an irrational drunken state rather than sober; their bodies are stooped and grotesque in appearance rather than upright and erect; their language is the broken dialect of immigrants rather than the proper speech of the bourgeois passengers above board; and their behavior is animalistic, aggressive, and instinctive rather than the sort of willed and deliberate behavior we expect of civilized men. In short, this hidden geography of the stokehole and the subaltern class associated with it function in the play as a raw underside to the public life of the middle- and upper-class passengers above board.[44]

Standing in stark contrast to this subterranean world of coal is the public promenade of the ship, located physically above deck, which finds expression in the play in the character of a young steel heiress named Mildred Douglass, a wan and privileged child of civilization. O'Neill portrays this world of the bourgeois consumer as leaching off of the raw, creative, and destructive energies of the stokehole. Whereas Yank represents the hidden, repressed, and motile energies below deck that drive the modern world, Mildred, who we find languishing in a lounge chair above deck, stands in for an artificial and anemic surface to that world. The play portrays her as "sapped . . . of . . . energy," "a waste product of the Bessemer process," and a mere by-product of the "artificialities that energy had won for itself in the spending."[45] Mildred personifies, that is, the world of the modern bourgeois consumer, a world blissfully ignorant of its dependency on these destructive physical powers in the boiler room and of its relationship to this subaltern class of coal stokers who generate that power.

The play's narrative is set in motion when these two halves of the modern world collide. In that collision, this repressed underworld of coal erupts violently to the surface and effects a rupture in the social and symbolic order of modern life. The key turning point in this respect arrives early in the play when Mildred, dressed in pure white, descends into the dark bowels of the ship to see its boiler room so that she can touch reality for a minute and feel the "thrill" of what one critic has termed a "vital contact" with the dirt, labor, and power of the furnace room.[46] But her descent into this subterranean space, like Beymer's previous descent into the coal mine, does not turn out as she anticipates. The experience proves terrifying rather than thrilling. As Mildred leaves behind the gentle air of the promenade to enter the hot heat of the stokehole, she finds herself in a hostile setting, surrounded by half-human figures "crouching" in the "inhuman attitudes of chained gorillas" and overwhelmed by the deafening din of "grating, teeth-gritting grind of steel against steel, of crunching coal" and the "roar of leaping flames." At the center of this dark world is the bestial figure of Yank standing "snarling" and "murderous" and wielding a shovel above his head in what appears to be an act of aggression toward Mildred.[47]

The trauma generated by this collision between Yank and Mildred connects us back to the psychic and social traumas that informed the middle-class relationship to coal in these years. For Mildred, the mere sight of the stokehole and these subaltern coal stokers—with their raw animality, their brute degradation, and their primal unrefined power—proves to be too much for her to bear. The experience leaves her, O'Neill writes, "paralyzed with horror, terror, her whole personality crushed, beaten in, collapsed, [defeated] by the terrific impact of this unknown, abysmal brutality." And although Yank does not so much as touch Mildred in this scene (in fact, her intrusion into his space causes him even greater trauma), the sudden knowledge she gains about his presence (and the presence of this dirty material underworld) leaves her feeling, in the words of one critic, "raped" by the experience. The "*throbbing* beat of the engines" and the "*quiver* of liberated energy" (italics mine) that serve as the backdrop to this event reinforce that traumatic interpretation.[48]

In this famous incident in the boiler room, O'Neill imagines a public reckoning with the social, psychic, and symbolic schizophrenia that shaped the modern landscape of power in the industrial age. Mildred's descent into the boiler room has the effect of raising this repressed underworld of coal—a world typically sequestered out of sight and buried in the depths of the modern psyche—to the surface for a moment, and the consequences of that confrontation are traumatic for everyone involved. Although the play does not tell us what happens to Mildred after she is whisked away from this world of the coal stoker, we do know that Yank himself feels psychologically emasculated by the experience, and that the self-knowledge he subsequently gains about his own positioning and degradation in the modern world sends him hurling into a self-destructive cycle of rage. Yank's decline reminds us, that is, that this traumatic narrative of coal sketched out in the previous examples always came back to the underlying fear that the white male body was being politically, economically, and socially emasculated in the world of cheap power.

But to the extent O'Neill's play turns on this original "rape" of Mildred, we can understand that event to be one that originates not in some *physically* traumatic encounter that she experiences in the stokehole but in a *symbolic* rupture in her worldview that comes from the violent and jarring knowledge she gains about herself in this confrontation—namely, that her own life derives from, rises out of, and is the cause of this other "abysmal brutality" in the boiler rooms and coal mines of the nation. This symbolic hemorrhaging in the boiler room speaks to the collective unease, repression, and even guilt that middle-class Americans have long had to this dirty material object that has such claims on our national history and modernity. And the inability of the play's protagonist, Yank, to find a voice or a place in the modern world after he emerges from its dark depths in the stokehole is an apt symbol of the unspeakability of the modern relationship to this barbaric facet of ourselves that continues to lie buried in the coal mines and boiler rooms of the world.

Only White Men Digs Coal There

This trauma in O'Neill's play is perhaps as fitting an ending to this chapter as we can find; yet there remains an obviously neglected chamber down which to descend in order for us to have a fully honest confrontation with this geography of power that fuels our world. Coal might have been (in fact, it surely was) an essential part of the story of modern class exploitation, environmental degradation, and psychic trauma that came with the transition to a fossil-fueled world, but it was also embedded in a hidden story of racial trauma as well; for in the hierarchy of injury and neglect characteristic of the modern production of power, few had it worse than did the African American coal miner who labored in an ethnically and racially stratified industry that left few doors open for anyone.[49] That is to say, even in spite of the relative racial equality and sometimes interracial solidarity to be found among members of the Western Miners Federation, the injuries associated with coal were always differentiated along racial lines, ensuring that modernity's power carried within its pulse the nation's long history of racial exploitation.[50] In such a context, the fact that Eugene O'Neill dared to cast the African American bass singer Paul Robeson in the lead role as Yank in a later remake of *The Hairy Ape* (in what was a daring inversion of the play's original racial subtext) is only further proof that the nation's primary discourse of coal concerned the blackening of white lives in the wake of fossil fuels.

In order to recover a glimpse of this racialized trauma that further problematized the nation's encounter with coal, this chapter draws to a close by venturing a bit further onto the periphery of the national imaginary. In the final two examples that end this chapter, we see surfacing —if in a marginalized way—the type of racial wounds that further perverted the modern geography of power. Both the first, a forgotten poem by Carl Sandburg about an African American coal miner named Henry Stephens, and the second, a low-budget foreign film on coal mining starring the aforementioned American singer and actor Paul Robeson, attest, if in different ways, to the disquieting and neglected racial unconscious that has also left its unfortunate imprint on the material and discursive history of fossil fuels in the United States.

Few readers saw it—but buried in the back pages of a special issue of *Survey Graphic* devoted to the subject of coal in the United States was a poem entitled the "Sayings of Henry Stephens" transcribed from interviews that Carl Sandburg, the poet, had conducted with an African American coal miner by that name back in 1917. Published on the eve of a nationwide coal strike that closed down the nation's coal mines for 163 days and that put more than half a million coal miners out of work was, in other words, this trace of an African American presence in the coal fields buried beneath the journal's main story of white miners' experiences in the mines.[51] In this marginalized text we see the racialized wounds of an African American coal miner from central Illinois (albeit as they were filtered through the pen of a Swedish American poet from the Midwest) surface for a moment in the bourgeois imaginary before retreating back out of sight.

Sandburg's poem was a powerful montage of aphorisms about life in the coal mines that he had transcribed from his earlier conversation with this black miner who had worked most of his life in the segregated mines of Illinois. Most of what Stephens had to say about coal and coal mining concerned the difficulties faced by those in his class and occupation regardless of race: "The cost of things is too high," he said. "They ought to be brung down/where they's more equal like/with other things. . . . We got to work to eat./And the scripture says:/'Muzzle not the ox that/ treadeth out the corn.'" Voicing the miner's pervasive fear of poverty and hunger and the haunting specter of not being able to feed one's family, Stephens's aphorisms evoked the type of class traumas that affected black and white miners alike whose lives were structured by what one critic calls an economic "morbidity" that defined the upper limits of a coal miner's opportunities as well as the baseline fears that kept him bound to his own exploitation.[52]

Yet Stephens experienced an added element of racial terror and insecurity that intensified the already hard experiences of the coal miner. Even in the supposedly free states of the industrial North, there was no confusing a white coal miner's job with a black coal miner's in the early twentieth century:

Springfield is Abraham Lincoln's town.
There's only eight mines out of twenty

In Sangamon County
Where the white miners
Let a negro work,

If I buy a house right next to the Peabody mine
That won't do no good.
Only white men digs coal there,
I got to walk a mile, two miles, further,
Where the black man can dig coal. . . .

United means union.
And union means united.
But they's mines runnin' twenty-five years
And the white men never lets the negro in.[53]

"Only white men digs coal there"—Stephens's broken vernacular reminds us that there was always a harsh race politics in the mines in addition to the burdens of class that saw white miners fighting to protect their modicum of privilege and power in this hard-pressed profession. What that meant for African American miners, of course, was that the scars of a coal-wrought modernity always cut deeper on this other side of the color line. In Stephens's muted anger and in his quiet racial resentment, we can capture at least a glimpse of the racially differentiated pains, insults, and injuries that were absorbed by African American bodies working in the bottomlands of modernity's geography of power and progress. Those added injuries only rarely surfaced in the bourgeois imaginary.

But rather than end on that sour note, this chapter leaves off with a note of symbolic agency, albeit an ambivalent one that went mostly unheard in the United States. It concerns a low-budget Welsh film entitled *The Proud Valley* that starred international celebrity Paul Robeson as a heroic coal miner named "David Goliath."[54] This "crude" and "stiffly played" film, as one contemporary critic described it, entered into American culture briefly when it screened at the Little Carnegie in New York in 1941,[55] but unlike its more popular counterpart, the Oscar-winning *How Green Was My Valley*, this other film about an African American coal miner never hit the historical radar and instead dropped deep into the forgotten depths of the national imaginary. As we will see, the reasons it did so had everything to do with race.

Robeson's film serves as a fitting conclusion to this chapter for a couple of reasons. First, like much of the literature on coal mining, its main storyline concerned the social damage and class traumas faced by coal miners in the early twentieth century. Like Henry Stephens, the miners in this film confront the type of iron-clad structural morbidity that plagued coal mining wherever it went: coal miners in the film fail to find work; they face eviction from homes when their wages come up short; and they struggle to be able to marry, to build families, and to imagine a future without the security of reliable employment. But they also confront an array of injuries peculiar to their profession. Destabilized mine shafts, raging fires, methane releases, depleted oxygen, dirty and unhealthy bodies, and finally death — all of these dangers portrayed in different episodes of the film link coal metonymically to the type of physical traumas we have seen in the other examples examined in this chapter. And although that connection is sometimes buried to make way for a more hopeful message about worker solidarity in the mines, it is made perfectly explicit in the film's traumatic climax when the main character, David Goliath, sacrifices his own healthy body to cover a mine explosion in an attempt to save his companions from certain death and in a specious subplot concerning the British war effort to finish the job at hand by replenishing the country's diminishing stock of fossil fuels.

In this climactic death scene, which ends the film — with fires raging in a side chamber, with oxygen in short supply, and with rock walls collapsing from a detonated charge — the tight symbolism that ties together coal, trauma, and modernity becomes overt. What we see in this scene is that the grisly death of a coal miner whose body is blown to pieces in the mine serves not only as the predicate for the survival of the rest of the white characters in the film but, in its subplot concerning Britain's dire fuel shortages in World War II, as the required sacrifice for the survival of the West against German barbarism. The main point here is thus a fitting one for this chapter: the ghastly destruction of a coal miner's body both defines the nature of coal's production in the film and, even more importantly, serves as the required sacrifice for the survival of western modernity. Coal and dying workers share close quarters, in other words, in this alternative film about fossil fuels and the fate of the modern world.

Although that part of the story could be told without reference to race, *The Proud Valley* makes race central to the story in a way that is out of keeping with the American discourse of coal. If the presence of black coal miners in the coal fields was well known to those who defended and contested the color line in their own lives, the broader popular dis-course of coal in the United States almost always elided, marginalized, or silenced their experiences in the public transcript. The net result was that even in the broken story of coal and coal miners, the narra-tive was almost always (as it was also with the stories of the Dust Bowl refugees) a story about the degradation and sacrifices that the nation's white underclass had made to support the wealth of a white middle-class nation and the maintenance of its modernity. But here in this B-quality Welsh film about coal mining stands this African American Christ fig-ure[56] whose heroic death attests not only to the historic class sacrifices in the mines but also to a racial trauma in the mines that was always part and parcel of—to quote O'Neill's protagonist—making the modern "woild move."[57]

The next chapter turns from this tortured relationship that Ameri-cans had to coal to the equally conflicted story of its sublimation as electricity.

4 | The Dynamo-Mother

A mingling of pain and loving consummation.
—Eugene O'Neill, *The Dynamo* (1927)

Somewhere on Broadway in 1929 a young man got down on his knees and started praying to an electric generator—to what he called the "Mother of Life—Big and Warm," or, for short, the "dynamo-mother." Confronted by an existential crisis that had taken over his life, this young man thought he had found the answer to life's questions in electricity—in that one big buzz that connected the universe. But troubled by the lonely hum of a godless existence and unable to get confirmation that he was right, he felt desperate to force himself to spiritual resolution. In what turned out to be a bizarre attempt to prove his fidelity to electricity, he pulled out a gun, shot his girlfriend, and wrapped his thin arms around the dynamo's hard metallic body in a weirdly literal embrace of its material energy. If he had hoped to find maternal warmth in the machine's "throaty metallic purr," that young man was, of course, sorely mistaken. Seconds later he lay crumpled on the ground, lifeless at the foot of the dynamo, in what the playwright described to be a grotesque "consummation" of man and electricity.[1]

Eugene O'Neill's play *The Dynamo* was a terrible flop by everyone's reckoning. It was pilloried by critics, avoided by audiences, and ultimately repudiated by the playwright himself. But as a cultural artifact, it returns us to a time when electricity held a deep discursive resonance in American life, that is, to a time before high-powered transmission lines receded into the background and became an unseen, albeit thoroughly embodied, part of the modern habitus. In O'Neill's play, we see, in other words, the nation's nearly universal tendency to fetishize this new energy of modern life while also getting a glimpse of its equally intense bouts of existential angst over the nature of modernity's energies and what possession of them was doing to its traditions, values, and spiritual health. Despite its poor popular and academic reception, O'Neill had

Praying to the generator: Eugene O'Neill, *The Dynamo*, 1929. Martin Beck Studio, New York. Courtesy of Imagno/Hulton Archive/Getty Images.

reason, in other words, to believe he had written in *The Dynamo* a "big work," one he called a "biography of . . . the American soul."[2]

This chapter, like the previous one, returns us to the troubled depths of that soul: first, by showing how electricity—which sublimated the presence of fossil fuels in American life—evoked our collective fantasies for a clean and tidy modernist sublime, or even industrial pastoral, that would join the high-energy society, without coal's messy implications, to the type of Jeffersonian dreams and republican norms of a bygone somatic age, and second, by recovering the persistent, if repressed, nervousness—or what amounted to a secondary type of trauma—that Americans had about living in this increasingly networked and centralized world of fossil fuels that seemed to be wholly owned, operated, and abused by a power elite that worked behind the scenes, that was immune to public scrutiny, and that was intent on using its control of the nation's energy and power for its own personal gain.

Steam Makes Slums

Unlike coal, electricity was never a dirty secret in American life. Whereas coal called up depressing images of toxic smoke, angered workers, and black slag, electricity erased the presence of fossil fuels on the consumer's end. By sublimating our dependency on prehistoric carbons and leaving just a soft purr at the endpoint of consumption, electricity was thus always aligned more naturally to what David Nye has termed the "technological sublime."[3] Americans consequently spoke of electricity almost universally in superlatives and archetypes. "Promethean" gift of the gods, "captive thunderbolt," "the white giant," "superpower," "giant power," the harbinger of a "white revolution"—electricity evoked the lyrical in the bureaucrat, it moved the fact-obsessed progressive to deal in myth, and it occasionally even inspired the playwright to write. The nation's collective embrace of electricity was, in fact, so intense, Carolyn de la Pena tells us, that for a short while we did some very peculiar things with it. We experimented with attaching it to our genitals, with wrapping it around our midriffs, and even with shooting it into our nervous systems in the hope that its life-giving force would rejuvenate modernity's tired and broken bodies.[4]

In such an enthusiastic context, O'Neill's failed meditation on electricity was an outlier, a depressing caveat to the nation's otherwise early love affair with this new energy.

Electricity in its modern form, as a source of heat, light, and work, entered into national discourse in a very specific way—in dialectical opposition to the declensionist narratives of coal and steam that had dominated the nation's collective meditations on the meaning of fossil fuels and industrialization. For both technocratic and political reasons, critics writing in the early twentieth century widely believed that the transition from steam power to electric power would have a revolutionary impact on the character of the nation's modernity. As General Electric (GE) engineer and maverick intellectual Charles Steinmetz put it, electricity prophesied a "white revolution" in modern life, a paradigmatic overhauling of its physical infrastructure (or what Thomas Hughes has called our "material constitution") to bring it more closely in line with the type of republican values and Arcadian fantasies that preoccupied the bourgeois imaginary well into the twentieth century and its age of power.[5] Electricity promised, in other words, to remake the material infrastructure of modern freedom.

Steinmetz was not alone in voicing these opinions about electricity. He was joined by a chorus of prominent actors on the political scene, including Gifford Pinchot, Stuart Chase, Lewis Mumford, Herbert Hoover, George Norris, Henry Ford, and many other intellectuals, activists, politicians, and industrialists, all of whom believed that electricity might be used to civilize coal's savage intrusions into modern life. The way Mumford put it was that electricity had allowed the West to supersede its "barbaric" industrial phase (what he called mankind's "paleotechnic" phase) and enter its more enlightened "neotechnic" one wherein modernity's power would finally be brought under human control and put into service of the public good.[6] Electricity would, one hoped, civilize coal's ragged edges.

These earlier observers were not just idle stargazers. Not only did electricity sublimate our dependencies on coal but it also altered the technological landscape of power. That is, earlier critics had some reason to believe the new energy technologies might affect the nation's culture complex in a beneficial way. Their hopes in this respect were pinned to the premise that most of modernity's social, economic, and environmental problems could be traced back to the technological limitations inherent in coal and steam.

Critics contended that the technological shortcoming of coal and steam was that it could not distribute its heat or power very far from the source of coal's combustion. Because steam power was distributed through an overhead arrangement of leather belts and metal shafts, the early steam-powered factories, they explained, had to be architecturally designed to bring workers close to the furnace, where coal's energy was originally released. Moreover, because the value of steam power declined rapidly the further one moved away from the source of combustion, this early energy regime required, in effect, the type of cramped factories, smoggy skies, and crowded cities that characterized US industrialization and that progressives found to be so lamentable. "Steam makes slums" was the way Gifford Pinchot put it. Like "a magnet drawing human iron filings" to itself, steam power lured people off of the land, he said, and drew them into "unhuman" cities and factories where the world's power and consequently its jobs increasingly resided.[7]

The term *centralization* served in public discourse as the cultural shorthand for this technological problem that critics associated with coal and steam. It also, however, denoted a second problem that went hand in hand with this mechanical shortcoming. Coal and steam, unlike the traditional prime movers—such as horses, oxen, and humans—that once drove the engines of industry and agriculture, came with prohibitively high start-up costs, which tended to "centralize" wealth and undermine economic competition. Because acquiring a steam engine and constructing the physical infrastructure to distribute its power was within reach of only a handful of capitalists in the industrial age, the new power technologies had placed control of modern power in the hands of the few; and even those few people who had acquired the financial resources to generate the capital outlays needed to set up shop, so to speak, had themselves depended on new undemocratic models of economic centralization (such as monopolies, trusts, and holding companies) in order to offset their original capital outlays.

What that meant for the social mechanism was that despite all of the benefits of fossil fuels—and there were many—they had come with the unfortunate side effect of undermining the more decentralized forms of power, property, and entrepreneurship that had grounded earlier republican dreams for a middle-class nation of independent producers.

Consequently, critics foisted (albeit not always fairly) the full slate of modernity's problems on coal and steam. They argued that it had undermined local handicrafts and artisanal production, that it had degraded economic independence, that it had denigrated the value of physical labor, and that it had led to such far-ranging social problems as a rise in vulgar materialism, broken families, licentious dancing, neurasthenia,

contemporaries characterized to be a new bloodlust for pugi-
he-man movies that reflected an era of brute power.[8] In short,
tics concluded that the shift to steam power had undermined
material foundation of the nation's liberty, equality, and happiness, which they imagined was more secure under a more decentralized
somatic regime.

The promise of electricity was, in contrast, that it would "decentral-
ize" the architecture of physical power and along with it redistribute
social power. Almost universally, critics made the case that because electricity's power, heat, and light could be physically distributed through
high-power transmission lines over a broader geographical range, it was
a potentially democratizing technology that might be used to redistribute modernity's energies and rebuild the infrastructure of western freedom. The ease of distributing literal power to more people on farms,
in middle-class homes, and in poor tenements seemed to put electricity more naturally in line with the nation's traditions of individualism
and independence. Critics imagined, therefore, that an electric regime
might realign the high-energy, highly networked modern world to the
type of long-held liberal, and even republican, values that traced back
to a more localized age of foot traffic and muscle power. "Life no longer
[needed to] be subordinated to steam" is how one journalist put it.[9]

Popular economist Stuart Chase developed this argument in a 1931
pamphlet entitled *The Promise of Power.* Chase was something of a popular expert on the nation's energy matters, having written widely on the
subject for magazines such as the *Atlantic Monthly,* the *New Republic,* the
Nation, and *Fortune,* and thus his writings picked up on the pulse of national discourse on the subject.[10]

In *The Promise of Power,* he repeated the common complaint that modernity's ills could be mostly traced back to the technological limitations
inherent in coal and steam and that electricity might reverse the tide of
modernization by luring workers back out of overcrowded cities and redistributing the architecture of physical power in the country. By redistributing the flow of power and with it the rationale for human migration
to the cities, electricity, Chase explained, would "blo[w] the population
broadly over the map" and thus break the causal connection between
modernity and urbanization. The nation was on the edge, it seemed, of
what his peer Pinchot called a "veritable back-to-the-land movement."[11]

The less obvious result of such a radical rewiring of the power grid
was that Americans would presumably leave behind the supposedly

harmful aural, olfactory, and tactile habitus of the modern city. Bourgeois reformers, such as Chase, imagined a return to a more wholesome life in the country where preindustrial sounds and rhythms might, as he put it, "react—favorably—on ears, eyes, nose, and skin cells."[12]

Chase played out this logic of decentralization by turning to his own personal experiences with the new electric technologies. The advent of electric heating, electric lighting, electric water pumps, and different electric motors, he explained, had made it possible for him and his wife to relocate from Boston to a quiet farm in Connecticut without foregoing the physical benefits of modern power and the leisure it implied. For the first time in history, he explained, rural Americans such as himself no longer had to live in the type of primitive and energy-scarce world that had previously all-but-defined life outside the cities. Once networked to the grid, Americans would have at their disposal, day and night, regardless of weather or season, not only reliable heat and light but also the physical power of electric motors that put them in possession of the equivalent of a full gang of hired laborers—"the equivalent of 26 men—200-pound fellows, number 19 collars, fists like California hams"—ready to perform most types of work on the farm. With such energy and physical power pulsing through the networked countryside, the nation's farmers would have access to a quantity of power that no American short of the southern gentry, with their hundreds of slaves and brutal disciplinary regime, had ever possessed.[13]

Electricity would, that is, allow the modern farm to get beyond the nearly iron-clad ecological and biological barriers to productivity. If in the somatic energy regime, the pace and scale of labor had been determined by an energy calculus dependent on solar power, plant photosynthesis, and the human and animal labor that could be fed during any given season based on the best agricultural practices of the time, the new energy regime, which gathered its energy resources from below the ground and delivered them by wire, augured a wholly different order of things on the farm. This new bioenergetic economy implied by electricity meant, in effect, that the nation might, if it chose, return to the type of Jeffersonian pastoral that populated its waking dreams.

The second argument Chase developed in his writings centered on what he saw to be the decentralizing impact electricity might have on US manufacturing. Whereas the coming of coal had done so much to concentrate industry in a few urbanized spaces and in a limited number of steam-powered factory sites, electricity, he argued, had enabled small-scale manufacturing to once again take root outside the current geographic centers of power and industry. Although Chase produced several examples to prove this point, his most compelling evidence turned on the experiments of automobile giant Henry Ford, who was in the 1920s and 1930s playing with the principle that electricity might be used to dismantle the central factory system and replace it with smaller-scale units of production spread out across the countryside.

While Ford was, in many respects, the nation's foremost architect of industrial centralization, he, like others of his generation, was deeply troubled by the social developments he had helped to set in motion. At his Highland Park plant (at one time the biggest aggregate factory in the world and an imposing symbol of what James C. Scott calls high-modernist architecture), Ford was experimenting in these years with using small-scale hydroelectric power plants to restructure his famous assembly lines. Having built a series of such power plants along Michigan's Rouge River, Ford dreamed of taking each of the central facility's different functions and moving them to individual sites along the river where he imagined employees dividing their time equally between working Michigan's rich soil and using electric machinery to complete industrial tasks such as steering wheel production currently being done in the Highland Park Plant.[14]

"I see the time soon coming," Ford said, "when the farmer will not only raise raw materials for industry" but will do the initial manufacturing of the product "on the farm." These strange experimental farm-factories, where workers divided their time between activities such as picking strawberries and manufacturing car parts, might have reflected an idiosyncratic, and decidedly top-down, dream for decentralizing production but, as Chase saw it, such experiments heralded a fundamental redistribution of the nation's productive capacity and a revolutionary transformation of modernity's social geography.[15]

The promise of electricity was not, however, simply restricted to dispersing persons and industry after the top-down manner proposed by Ford; it was also about liberating and empowering people in a world in which individual agency had, in some respects, been so drastically diminished. Electricity's boosters imagined that electricity might literally put socioeconomic power back into Americans' hands by providing all Americans with the functional equivalent of the teams of horses and workers (i.e., horsepower) that they needed to compete more effectively in an industrial marketplace defined by a scale and pace of work that exceeded the capacities of any individual body. Drawing explicit analogies between electric motors and human bodies—those *nineteen-inch collared fellows* Chase imagined working for him when he flicked the switch of a motor—electricity's proponents imagined all Americans having access to a degree of physical power that would serve as the material foundation for both a renewed individualism and a renewed economic independence.

As its boosters knew, there were, of course, no guarantees electricity would lead to this type of decentralization and democratization of industry, people, and socioeconomic power. Although they imagined it might be used to level the playing field in a world structured by such vast inequalities of physical and social power, they were also wary of adopting the sort of technological optimism that had originally attended the birth of coal and steam. Having witnessed the tortured path down which a coal-fired benevolence had travelled, these early observers foretold electricity's future in question marks rather than in confident conclusions. Did electric power mean an end to the "curse of . . . scarcity," the growth of a "new regionalism," the "collapse of . . . out-grown theories of class struggle," the realization of "universally high standards of living," the proliferation of "jobs, leisure, freedom," and the end of "drudgery, congestion, noise, smoke, and filth"? Or would electricity, they asked, merely intensify the patterns of social and environmental exploitation initiated by coal and steam and make permanent what those earlier technologies had thus far begun?[16]

"This is the Day of Choosing," journalist Joseph Hart wrote.

> We stand, today, where the Greeks once stood: face to face with Fate. We have Power beyond their dreams of power: power that indisputably belongs in the realm of nature, the proper use of which need not degrade a single human being. . . .
>
> We can build communities upon the foundations of great but decentralized power. . . . We can provide the materials out of which men can make for themselves the manner of life they prefer. Or, we can surrender to the control of the greater machine; permit electricity to make permanent what the steam-engine began, be happy in the roar of industry and lose all . . . sense of freedom, justice and beauty.[17]

Worrisome words for these otherwise technological optimists.

Superpower and Giant Power

Americans' faith in the democratizing potential of electricity remained high throughout the twentieth century. Although coal persisted in the popular imagination as the poster child of industrial trauma, electricity typically represented to Americans their chance to have a second go at industrialization, a second chance to *rewire* modernity, so to speak, in order to redistribute its literal powers and democratize its energies.

This therapeutic potential inhering in electricity came, however, with a caveat. It hinged, critics said, on how electricity was to be culturally appropriated and politically administered. The high stakes perceived in electricity's future had, of course, direct implications for electricity's life as a signifier. As proponents of private energy development and advocates for public control lined up on opposite sides of the political aisle to stake out their claims over the new energy regime, electricity moved to the symbolic center of a hyperbolic culture war over the meaning of that power and the role of its energies in modern life. The high-pitched rhetoric that characterized that earlier discourse of electricity can be glimpsed in the Superpower and Giant Power debates of the mid-1920s as Americans debated the future of the nation's electric development.

The term *Superpower* was the original term in this debate. Derived from a federal energy survey conducted by the US Geological Survey (USGS) in 1919, it referred to a federal plan for integrating the nation's power-generating capacity into a single superpower grid in the industrialized Northeast not unlike what we have today. Born out of concerns raised during the wartime fuel crisis, when coal trains failed to deliver coal on time and when coal shortages forced consumer cutbacks, this dream of a superpower zone aimed to eliminate waste and inefficiency in energy production and to ensure consumers and industrialists a more reliable supply of energy in the heavily populated Northeast. This idea behind a superpower zone reflected a conservationist consensus on both sides of the political aisle that assumed the need for pooling electricity between Boston and Washington into a single coordinated grid, developing the region's remaining hydropower, and rationalizing the region's railway transport system for coal.[18]

But because the idea of a superpower zone was rooted in private energy development with no assurances for public regulation or control, critics of the scheme associated it with what they saw to be an already dangerous tendency toward consolidation of the nation's energy infrastructure in the hands of the few. Critics were not far off on this score. Because proponents of the superpower scheme required the industry's cooperation, they had promised at the outset that the region's energy infrastructure would be controlled "lock, stock, and barrel," as one of them put it, by private interests. Superpower's critics seized on that fact and quickly coopted the term to designate the seemingly superhuman concentrations of private wealth and power that such a system implied.[19] In fact, the nation's largest third party, the Progressive Party, subsequently made public control of hydropower and coal a defining feature of its national agenda.

In popular discourse, the figure most closely associated with this concept of superpower was industry magnate Samuel Insull, who had amassed unprecedented financial control over Chicago's energy infrastructure and that of the surrounding Midwest and who liked to refer to his own holdings in such grand ways. Appearing twice on the cover of *Time* magazine (first as a capitalist titan and second as a disgraced villain) and showing up in bit roles in the nation's novels, newspapers, and plays, Insull became in the popular press the personification of what contemporaries at the time called the Power Trust, that small coterie of private energy financiers who owned and operated the nation's electricity.

We can get a glimpse of the hyperbolic rhetoric surrounding this term, *Superpower,* by turning briefly to a text such as *The Big Money,* in which novelist John Dos Passos caricatured in a sort of impacted and dizzying prose the rise and fall of this man he called by the absurdist name "Power Superpower."

Power Superpower

. . .

The Insull companies spread and merged put competitors out
of business until Samuel Insull and his stooge brother Martin con—
trolled through the leverage of holding companies and directorates
and blocks of minority stock
light and power, coalmines and traction companies
in Illinois, Michigan, the Dakotas, Nebraska, Arkansas, Oklahoma, Missouri, Maine, Kansas, Wisconsin, Virginia, Ohio,
North Carolina, Indiana, New York, New Jersey, Texas, in Canada, in Louisiana, in Georgia, in Florida and Alabama.

. . .

He held directorates in eightyfive companies, he was chairman of sixty-five, president of eleven: it took him three hours to sign his resignations.[20]

This farcical (if worrying) notion of a power so vast one man could leverage control over more than a hundred different energy companies and spend three hours simply signing his own resignations from those companies after bankruptcy and allegations of fraud was an alarming one that made for good narrative in the populist 1930s, but for more sober critics it was simply the "greatest industrial fact" of the time.[21]

Standing opposite the notion of privately administered Superpower was this other equally hyperbolic term *Giant Power*, which circulated as the progressive alternative to private energy development and which was tied to the idea of public control, or even ownership, of the nation's energy infrastructure, including its coal plants, its hydropower, its electricity, and its fuel delivery systems. With an etymology tracing its own origins back to a similar state-sponsored energy survey commissioned by the coal- and oil-rich state of Pennsylvania, Giant Power added to the conservationist principles of the Superpower agenda a democratic premise that all Americans should control and have equal access to the energies of modern life and the physical power they provided.

Closely associated with the state's conservationist governor, Gifford Pinchot, who had authorized the study, the technological core of the Giant Power plan centered on a proposal to relocate the state's dirty coal plants to the mouths of mines in the hilly western part of the state, where fuel was plentiful and people scarce, and then deliver that energy back east by high-tension transmission lines. The plan also included a conservationist scheme to capture by-product waste from the combustion of coal and reuse it in manufacturing plants, to develop the state's remaining water power, and to push for increased regulation of the utilities. But its real cultural substance concerned the strident moral assertion that redistributing and expanding access to material power would both physically and socially re-empower those Americans—especially rural farmers—who had been left behind by the modern age.[22] Anticipating the later New Deal programs for rural electrification in the 1930s, Pinchot and his peers imagined a fully electrified countryside integrated into the modern grid.

This rhetoric of Giant Power was an aggressive one. It cast the terms of debate over electricity in Manichean terms. On the one side, Pinchot wrote, was a "friend and fellow-worker of man," a "beneficent power" that would democratize energy resources, liberate Americans from the pains of industrial and preindustrial work, and return to them a degree of social power they had lost in the transition to industrial modernity; on the other side, he wrote, lurked an "evil power," a privately owned and secretly run energy system governed by a cabalistic Power Trust. That trust, he wrote, operated behind the scenes like an "evil enchanted spider . . . hastening to spread his web," and it reached out "to control and live upon the life of our people." Not unlike the contemporaneous debates over the steel, beef, and tobacco trusts, this high-pitched rhetorical contest over private energy development framed the public discourse of electricity within the polarized terms of freedom and slavery, good and evil, commonwealth and wealth, public and private power.[23]

In that heated context, public power meant more than simply economic control. Power was embedded in a much larger narrative concerning the potential regeneration and emancipation of society in which electricity was to be a "civilizing" influence on a reckless industrial world characterized by what Lewis Mumford called "giantism" and "crude" uses of power. In other words, control of electricity was an important issue for progressive critics precisely because it meant to them the possibility of a better world—one where the "smoke pall of paleotechnic industry begins to lift," where the "clear sky and the clean waters of the [earlier] eotechnic phase come back again," and where modern power finds "higher refinement" as it reaches out to a new "integration" with life.[24]

Admit Your Sins and Repent

That ethical discourse of power was yet in its formative stages during the Giant Power debates of the 1920s. But by the 1930s—when the Power Trust came under official public scrutiny and when Depression-era exigencies allowed for comprehensive energy reform—it had grown to full maturity. Federal Trade Commission (FTC) investigations into the utilities' practices, the announcement of high-profile energy programs such as the Tennessee Valley Authority (TVA), Bonneville Power Administration (BPA), and Rural Electrification Administration (REA), and a steady stream of New Deal propaganda touting the virtues of publicly controlled power all worked to push electricity from the margins of the national imaginary to its center.

What arose from these later developments was a moralistic discourse of power that defined public control of electricity as the essential building block for a modernized, high-energy republic in keeping with the sort of republican traditions and liberal values that traced back to an earlier age of horse carriages and yeoman farmers. In effect, electricity appeared in this discourse as the means for civilizing the machine that had grown to such monstrous proportions in an earlier age of coal.[25]

The 1928–1931 FTC hearings on the utilities might seem like an unlikely place to turn for an inquiry into this popular discourse of power, but the findings of the FTC wended their way into public discourse through a 1932 muckraking exposé, *Confessions of a Power Trust*, written by a cofounder of the Public Ownership League, Carl Thompson, a man who had been hounded by the utilities for promoting the idea of public rather than private power. These so-called confessions of the electric industry raised to the surface a morally corrupted and privatized world of power where backroom deals were daily eroding the democratic potential of the new energy technologies. As Thompson saw it, the pressed coats and gilded doors that marked entry into this other world of power were simply the disarming veneer for an equally ugly penthouse counterpart to the traumatized mining regions.

Electricity's critics described this other world of power through a language of repression, confession, and revelation that denoted a secondary type of trauma. In this case, trauma concerned the shocking revelation that a so-called Power Trust was acquiring through unscrupulous means nearly complete control over one of the Malthusian essentials of life—the nation's fuel—and that it was, moreover, using the financial clout it had gained through such means to rescript the public debate over power and to shield itself from scrutiny. As public power spokesman Senator George Norris put it, "The country will be dumbfounded and amazed when it learns that practically everything in the electrical world is controlled either directly or indirectly by this gigantic trust . . . [a] trust that controls from one end of the country to the other the generation and distribution of electricity . . . running all the way from a little electric bulb . . . to the gigantic generator that will handle without trembling from 30,000 to 60,000 horsepower."[26] Reinforcing that point, Pinchot had suggested that "nothing like [it] has ever appeared in the history of the world . . . [or] remotely approaches it in the thoroughgoing, intimate, unceasing control it may exercise over the daily life of every human being within the web of its wires."[27]

This moralistic discourse of power, which traced itself back to the Giant Power debates, drew a direct connection between physical and social power by turning attention to the disturbing fact that the nation's electricity, including its generating capacity, its distribution systems, and its commercial technologies, was wholly owned by an increasingly small number of financiers who conducted their business behind closed doors. The novelty of the FTC investigations, Thompson said, was that they had pried open this private world of power for the first time by cutting through its constructed "labyrinth of secrecy . . . and . . . designed confusion." In effect, the FTC hearings gave Americans access for the first time to the industry's internal workings by opening up its private account books and correspondence so that the public could get a glimpse of what Thompson called the industry's "modus operandi" and "innermost methods."[28]

Much of Thompson's book was given over to the type of tedious financial matters that characterized, and still characterize, the financing of power. The devil was, however, he argued, in the arcane financial details that only appeared to be beyond the ken of the ordinary citizen. Behind the mask, things were not nearly as complicated as they seemed and looked more like straightforward robbery of the public trust. Strategies such as using interlocking directorates, scaffolding multitiered holding companies, dividing privileged voting stock from nonvoting stock, and watering down stock to inflate its values had all worked, he explained, to concentrate individual decision making and industry profits in worrisome ways.

Summarizing the FTC's findings, Thompson explained in individual chapters, for instance, how relatively small financial investments in limited-voting stock could allow a smart financier to acquire nearly total decision-making power of a heavily capitalized company; how interlocking directorates (wherein an individual investor directed multiple companies that were often connected through a single higher-level trust) could allow a player in the industry to control so many different subsidiaries that he found himself in the lucrative position of negotiating contracts with himself; and how the use of less ethical mechanisms such as inflating stock values and differentiating company earnings between different types of stock could help a savvy investor to circumvent the profit limits set by public utility law. The fact that twenty companies and their twenty boards of directors already controlled about 60 percent of the nation's commercial power in the 1920s was, Thompson thought, perhaps the most important economic fact of his time.[29]

The big "sins" of the Power Trust, however, did not simply turn on concentration of ownership but on how the industry had used its financial clout to cover up systematic subversions of public utility laws and to distort public discourse in such a way as to shield the utilities from open scrutiny. The surprise (and revelation) in the hearings was that even insiders of the industry were saying, albeit behind the scenes, what critics had long suspected of the industry—that the utilities were systematically and intentionally bilking the public. That was the hearing's most traumatic finding.

Thompson made much of the confessional language found in the private correspondence of officials in the industry who had expressed in their writings a "disgust" over what they saw to be the systematic "lying, trimming, faking, and downright evasion of trust" that supported the industry and the accomplishments of its "so-called big men." "Admit your sins and repent" one insider wrote to another: "What profiteth . . . a man to gain the whole world if he loses his own soul?" said another—and a third was found complaining he could not sleep at night and that one day he was "going to get a piece of soft pine and . . . whittle out a conscience that will let me sleep as other people do" before it was too late. The whole industry, it seemed, was populated, one man said, by a bunch of "yes ma'ams" who dared not speak out for risk of losing their jobs. The recent suicide of one of these anguished middle men struck Thompson as a damning indictment of the entire industry's guilty conscience.[30]

But the traumas to American society ran deeper than that. The high anxieties of a few middle managers paled in comparison with the shocking discovery that the electrical industry as a whole had been systematically working behind the scenes for years, as one industry official put it, "to change and direct the economic thought and economic practice of the American people." The FTC had found evidence, in this respect, of a staggering industrywide campaign to manipulate public opinion in favor of private power while distorting the facts about public power and the costs of state regulation. As one industry official bragged, the utilities had, as early as the 1920s, tried every known method of propaganda to remake public opinion short of skywriting. "I don't believe we have tried that with airplanes," he told the committee.[31]

The FTC's findings shocked even the most jaded in this respect. As Thompson explained, the industry's influence penetrated deep into the nation's educational and media institutions and included such unsavory practices as purchasing local newspapers to guarantee favorable coverage in battleground cities; pressuring college professors and publishers to rewrite textbooks so that they were friendly to private power; urging local businessmen to denounce public power spokesmen or even force cancellation of their speeches when they arrived in town; distributing proindustry curricula to schools and colleges to naturalize the notion of private power; contracting with leaders of women's clubs, churches, farm organizations, the Boy Scouts, the Girl Scouts, and other civic institutions to cultivate support for private power among women, youth, and farmers; organizing grassroots tea parties in homes to get mothers to buy into the notion of private power; relying on friendly news reporters to publish industry-produced stories that distorted the facts of public power; and funneling profits (paid by consumers) into industry films, radio broadcasts, billboards, flyers, leaflets, books, and advertisements that aimed to discredit municipally and state-controlled power. Thompson explained that such prolific cultural activity, coordinated through local, state, and national chapters of the National Electric Light Association (NELA) and the utilities' individual public relations departments, meant that the industry had set as its goal nothing short of revising the meaning and language of the nation's power.

In this campaign to manufacture consent for private power, the irony, Thompson said, was that the public itself was footing the bill. Because allowable profits for the utilities were pegged by law to 6–9 percent of a company's capital investments and operating costs (plus other outlays in any given year), the more money a company spent on propaganda actually raised that company's expenses and thus its potential earnings as surely as did investments in research and development. In the absence of a vibrant free market in electricity and without what contemporaries called a public yardstick to gauge the true cost of electricity, the utilities were able to fold an expensive culture war back into the costs of operation and then pass those costs onto consumers in the form of higher rates. As one NELA director told his peers: "Don't be afraid of the expense. The public pays the expense." What choice, after all, did a consumer have when faced with legally authorized monopolies on the market?[32]

Confessions of a Power Trust hit the reader over the head on this point. It described in detail the vast cultural apparatus the industry had developed by the 1920s to distribute its propaganda in schools, magazines, films, newspapers, textbooks, and homes. In the realm of news coverage alone, Thompson explained that the industry had gone so far as to purchase local newspapers to ensure favorable coverage, to cultivate private reporters who could be relied on to publish handpicked stories, and to use the industry's vast advertising muscle to weaken the resolve of commercial papers in their coverage of local power debates. In one egregious example, an operator in northern Illinois, Ira Copley, bought several newspapers in the state and then contracted with another media giant to purchase fifteen more papers in Illinois and California where his investments lay. The result of such control over the media meant that in certain parts of the country the debate over private power simply vanished from the public transcript.[33]

Added to such strong-armed tactics was the influence that could be bought with an advertising budget that amounted to as much as $25,000,000 annually. Not only did newspaper outlets refuse to bite the hand that fed them but the industry's financial clout allowed it to reach deep into the civic sphere, with its fingers dipped into private organizations like the General Federation of Women's Clubs (such as when it hired that group's president to write pro-industry pieces for *Better Homes and Gardens* and *Women's Home Companion*); the Boy Scouts (such as when it published "Reasons Why Public Utility Officials Should Become Active in the Affairs of the Boy Scout Organization"); college professors (such as when it debated the merits of funding teachers and professors in the "starveling" professions to nurture their support); and authors and school boards (such as when it coerced publishers, professors, and colleges to alter the content of textbooks and curriculum to support private rather than public power).[34]

In short, the traumatic truth about private power was that the utilities were not only remaking the nation's material infrastructure but also remaking the nation's superstructure by overhauling its culture and its politics. To quote historian David Nye, thousands of pages of testimony had revealed to the public that the industry was engaging in a "covert" campaign to remake public opinion of electricity through the systematic dissemination of "half-truths and at times outright lies."[35] In such an unhappy context, stricter regulatory laws favoring consumers, the expansion of consumer cooperatives and municipally run power plants, public oversight of holding companies, or even a more equitable distribution of electricity's benefits fell victim to the concentration of power and its influence. The result of that concentration of both material and social power was that our public discourse on the subject shrunk down in the United States to a simple and inane choice between the presumably good belief in loosely regulated energy markets and the treasonous beliefs in public power and strong regulations.

The Eagle Fades to the Octopus

Confessions of a Power Trust called up a secondary type of trauma underlying the nation's shift to the new energy regime, one that concerned the tightening relationship between physical and social power, but it, of course, failed on its own to dislodge the dominant discourse circulated by the utilities and backed by a long tradition of laissez-faire liberalism rooted in an earlier age of muscle power. Change did come, however, but it came from an unexpected quarter.

The nation's sudden plunge into a decade-long Depression in the 1930s did for the discourse of electricity in the United States what no federal report or muckraking text could ever have done. The unanticipated failure of free-market capitalism and the subsequent birth of the modern welfare state in the 1930s eroded the industry's influence on this discourse and replenished (at least for a time) what was becoming an impoverished vocabulary for speaking about energy and power in the United States. As historians such as David Nye, Ronald Tobey, Neil Maher, and Finis Dunaway have all pointed out, this paradigmatic shift in the nation's politics saw electricity as a cultural object migrate rapidly to the center of the national imaginary.[36] The emergence, in quick succession, of high-profile energy programs such as the REA, TVA, and BPA wrested a measure of control of electricity back from the private sector and opened the door to an alternative vision for an electrified world propped up not by the free market but by the state.

This revitalized discourse of energy and power revolved metonymically around triumphalist images of reinvigorated farmers, emancipated housewives, and clean and well-lit electrified homes each signifying in their own way Americans' emancipation and empowerment by wire. The populist tenor of that discourse was promulgated by a vast state cultural apparatus that included the output of any number of government agencies, including Farm Security Administration (FSA) films like *Power and the Land* and *The River*, Works Progress Administration (WPA) murals like *Progress of Power* and *TVA Power*, REA photographs and poster boards like Lester Beall's "Here It Comes," and BPA songs like Woody

Guthrie's "Roll on, Columbia" and "The Biggest Thing Man Has Ever Done." The type of rousing visual montage seen in a film such as Pare Lorentz's *The River*—with its blasted mountaintops, its dammed rivers, its pulsing wires, and its emancipated people—was simply the aesthetic climax to this emergent narrative of power, which tied regional regeneration, national equality, and physical emancipation to public control of the nation's untapped energy resources.

A final example, a government play titled simply *Power* written by Arthur Arent for the WPA Living Newspaper Project, brings us full circle to the tensions we have seen between the Promethean promises of electricity and the public fears that came with the monopolization of electricity by private investors. Performed in New York City in February of 1937 at the height of public power debates over the legality of the TVA, this government-funded play defended the notion of publicly controlled electricity by promoting the opportunity that it presented for liberating Americans from the uncertainties of their preindustrial past and from the economic dangers of its present.[37]

The subject of power was not subtle in this production. The play dove straight into its topic by projecting the word POWER onto the opening stage curtain and then flashing that word progressively larger and larger until the curtain finally opened. "Give us a demonstration of *Power!*" the narrator boomed out at the start of the play to which a flurry of multi-ethnic voices could be heard proclaiming the importance that electricity held to their work and well-being. We promptly hear a Jewish theater owner tell us that electricity turns the reels of his films and projects the light needed to run his talkies. We hear an Irish housewife cry out that electricity's heat is all that stands between her newborn baby and the cold blasts of an eastern winter outside her walls. We hear an Italian restaurant proprietor explain that electric lighting extends the hours of operation that make his business profitable. And we listen to the Anglo narrator rattle off the manifold ways in which electricity pulses through our lives, from the way its energy drives the looms of modern textile production to the more subtle ways in which it powers the fields of medicine, radio, telecommunications, traffic control, and food production.[38]

But as the play makes clear, access to electricity had also introduced into modern life new economic and technological vulnerabilities that our predecessors did not have to face. To dramatize these novel dangers, the play cut abruptly to an electrical outage (signaled by a loud boom offstage) that unexpectedly interrupts the dialogue of the play and casts the theater into complete darkness. Taking the script from the week's local news headlines, the play's narrator reads aloud the following lines: "December 28th, 1936: Newark, New Jersey, and its suburbs were thrown into total darkness tonight when fire in a power plant cut off all electric current. Nearly a million people were affected."[39]

This simulated blackout, which occurs onstage, introduces into the play a collective trauma. Following the loss of electricity, we hear a baby with the flu cry out in the biting cold as her mother frets ever more loudly about the loss of heat; we hear police sirens blaring while a traumatized driver can be heard muttering "I didn't see her. I tell you. It was dark. I didn't see her"; we see primitive flares light up on the stage as air-traffic controllers try frantically to guide pilots who can no longer see the city's landing pad; and we hear a number of equally worried voices screaming "Operator, operator" as they call out in the dark for help ranging from the relatively minor concerns of city bakers at risk of losing a day's profits in the ovens to the more worrying cries of nurses and surgeons scrambling to confront life and death situations without the type of light, heat, and power on which modern hospitals depend.[40] The point of the scene is that electricity is not simply a modern convenience but the material infrastructure of our work, our pleasure, and, above all, our security, the assumption being that the modern social order risks collapse without it.

Although Arent's play grabbed the audience's attention with this traumatic notion of living in a world without electricity, what made that message so powerful was the growing assumption among its urban audience that electricity and the power that it provided marked the real difference between modern and primitive life. The play dramatized this stark difference between a secure and modernized world of electricity and its impoverished and insecure pre-modern counterpart in its descriptions of the rural South. In parts of the United States, the narrator explained, Americans were still slogging out their lives in a "primitive, a throwback to an earlier America" without the advantages of modern power.

Parts of seven states, 40,000 square miles, two million people. All living in a region blighted by the misuse of land and by the wash of small streams carrying away the fertile topsoil. In these cabins, life has changed but little since some pioneer wagon broke down a century ago, and for them this became the promised land. Occupations—when they exist at all—are primitive, a throwback to an earlier America. Here stand the results of poor land, limited diet, insufficient schooling, inadequate medical care, no plumbing, industry, agriculture or electrification![41]

From environmental degradation to child malnutrition, from poor health care to lack of running water, Arent's play (and the larger discourse of power of which it was part) connected the absence of electricity to the type of poverty and ecological decay found in the nation's most traumatized rural regions, such as in the Tennessee Valley. Without electric pumps to deliver fresh water, without electric heaters to reduce the backbreaking work of chopping down trees, and without electric washing machines to liberate farm wives from the drudgery of hauling water and scrubbing clothes, much of the nation, the play suggested, was yet struggling to survive in a preindustrial world of overtaxed bodies and material want. The emancipating promises of electricity had, in other words, been squandered under private control and the profit motive.

To justify its support for public energy programs, Arent's play made clear that the sort of suffering implied in this portrait of the dispossessed was not a technological or Malthusian problem. Coal was not running in short supply, and the advent of alternating current had made it technologically possible to provide every American with nearly unlimited heat, light, and horsepower. The real threat, as we have already seen in the previous example, was an economic one that concerned private consolidation of the nation's electricity in a few hands and their subsequent misuse of that power. It was hard for the audience to miss that point as the message was broadcast loud and clear on stage in a crudely transparent transition in which the image of a proud American eagle, the sign of a robust nation, was projected on the scrim and then replaced by the image of an octopus, the period's sign for economic strangulation by monopoly.

This threat that an oligopoly on the nation's energy markets posed to a fully emancipated country was played out in several different scenes detailing both the various abuses of the Power Trust and its methods for monopolizing control over decision making and profits in the industry. In one wonderfully didactic scene, actors were featured moving large geometrical blocks around the stage (each representing a different holding company or power-generating facility) to show visually how a handful of private energy financiers had managed to create tiers upon tiers of fictional corporate identities that had the effect of centralizing profits and decisions in the industry at the top. That scene, which ended with the completion of a large pyramid onstage, illustrated how the use of these fictional identities (combined with loose regulatory laws and the division of company ownership into common and preferred stocks) worked to allow people such as Samuel Insull to acquire direct control of, in his case, more than a hundred different power subsidiaries. To capture the ludic nature of that concentration of wealth and power, the play imagined a moment in which Samuel Insull holds negotiations with himself (as the director of two different companies) simply by taking money out of one pocket and putting it into another while negotiating favorable contract terms with himself.[42]

Such consolidation of electricity in the hands of the few had as its complement the loss of agency for ordinary Americans. In this moralistic discourse of power, the average American was an increasingly beleaguered individual beset by technological and economic forces beyond his control and no longer sure about his own ability to affect the outcome of his life. The play personified that disempowerment in the twin characters of the meek consumer and the sheepish farmer, both of whom walked through the play feeling utterly disfranchised and powerless in a newly electrified world from which they had expected so much.

Both the consumer, who wears a C across his chest, and the small-town farmer, who was otherwise a hero in 1930s literature, are manipulated throughout the play and ultimately cheated by the industry out of their rights to cheap and affordable heat, light, and power. We see their disempowerment played out in various scenes. In one scene, for example, we see a timid farmer daring to approach a corporate executive with the request to run lines to his farm, only to get rebuffed and entangled in a web of deception. In another, we see a confused consumer appeal for lower electric rates only to discover he is trapped in a legally authorized monopoly and inadequately educated to understand the complexities of holding companies. And we are all forced to stand by and watch as politicians in the play are bought out by private energy interests, as college professors sign sweetheart contracts with the industry, and as loose regulatory laws shield the industry from having to make transparent its dealings. The overall impression is that electricity's promise of bringing empowerment and independence to ordinary people has been undercut by a web of new material dependencies and the structural disadvantage of living in a world of privatized energy markets.[43]

The real fulfillment of modernity's potential comes, as we might expect, in the form of state-financed power. The play, in fact, works hard to

Poster for the Living Newspaper production of *Power*. Blackstone Theatre, Chicago, 1938. Courtesy of Music Division, Library of Congress.

make its case for public power and to preempt whatever lingering concerns the audience might have had that state-owned power was somehow an overreach of the federal government's constitutional role (as the industry had argued). That message is developed in the play through an especially heavy-handed dialogue between a young girl and her father, who discuss aloud the validity of proposed New Deal power projects such as the TVA. As the girl prods her father (a classic liberal skeptical of public power) about the difference between government-delivered mail, which no one seems to question, and government-delivered power, which is being hotly contested, the adult character finds himself squirming intellectually to explain to her why the federal government should be charged with the essential task of delivering the post but should keep its hands out of the equally essential business of providing Americans with heat, light, and power. In a rebuke to her father's logic and to the false consciousness of liberals, the scene ends with her complaining to her dad that when it comes down to it "the people are awfully dumb" concerning their own self-interest and their susceptibility to corporate propaganda.[44]

This WPA play drew to a close with the haunting specter of nine skull-like faces, representing a conservative US Supreme Court, hovering over the stage in anticipation of the Court's decision on the TVA. As of 1937, when the last production of Arent's play was performed, the Supreme Court had not yet rendered its decision on the act's constitutionality, and thus critics could still imagine this battle between private property and the public welfare ending in some sort of modernist nightmare. Of course, when the Court did issue a decision, it came down affirmatively on the side of the project's constitutionality, which ensured that electricity, as Tobey has explained, would become a basic right of the modern welfare state rather than a privilege to be meted out according to market forces.[45] But that was never a foregone conclusion, and it generated at the time decidedly high public anxieties about private power's resistance to the spread of public electrification at the federal, state, and municipal levels.

In sum, this WPA production can serve for us as an example of the broader moralistic discourse of power that captured the public imagination in the 1920s and 1930s and that tied the idea of public power to the sort of comfort, security, and physical emancipation that a conscientious national energy policy could presumably guarantee. In this most memorable of Living Newspaper productions, we can see, that is, a concerted effort on the part of the state to reconcile the networked, high-energy

The Supreme Court decides the fate of public power: Arthur Arent's *Power*. Ritz Theater, New York, 1937. Courtesy of Music Division, Library of Congress.

world of coal and electricity to the sort of democratic values, liberal-individualist norms, and even republican virtue that came from an earlier age that turned on somatic energy sources. After all, here was the state using electricity to call into life a community of emancipated citizens liberated from the pains of preindustrial life, in possession of a power undreamed of a generation before, and rallied around the notion of the public good rather than economic individualism.

I Don't Want to Know the Truth

It would be too simple, of course, to end this chapter on such a sanguine and triumphalist note even if the nation's energy policy in the 1930s did indeed play a pivotal role in bringing power to the people, so to speak, and even if, for a short while, it appeared as though we were moving toward a more deliberately formulated energy policy on the federal level.

From the start, there were always, as critics had worried, obvious signs that Americans lacked the ethical foresight and political will to maintain their hold on the rambunctious energies of modern life and to steer those powers to enlightened ends; and that was true whether electrification was to proceed under private or public auspices. When in 1891 the New York criminal justice system electrocuted its first death-row inmate in a gruesome and botched death by electricity that promised to be the epitome of enlightened capital punishment,[46] when in 1903 Edison's General Electric publicly electrocuted a series of animals, including an elephant named Topsy, which it shot through with a 6,600 high-voltage charge multiple times, in order to prove that its competitor's use of alternating current was more dangerous than the company's own reliance on direct current;[47] when the future President Herbert Hoover dreamed of moving the nation's waterfalls simply to make them more efficient energy producers;[48] and when even the TVA—that model of democratized power—recorded some of the worst episodes in electricity's environmental history,[49] it was fairly clear that public and private power shared responsibility for the darker side of modernity's light. In other words, well before our consumption of electricity resurfaced in the gruesome forms of climate change, mountaintop removal, mercury-impaired infants, and spoiled drinking water, we were already struggling to deal with a physical power that had outstripped our collective comprehension.

"I don't want to know the truth! I just want you to hold me, Mother." Those words rang out at the end of Eugene O'Neill's play just moments before his main character grabbed onto the dynamo in a warm and loving embrace and electrocuted himself.[50] Given both the depths of our own dependencies on electricity and the sublimated costs of those hard material dependencies, such a loud denial—"I don't want to know the truth! I just want you to hold me"—is a fitting ending to this moralistic discourse of power. For in it we hear what we might call the echo of a nervous underside to the modern self that remains troubled by its worrisome material dependencies even as it insists on being ever cheerful, even sprightly, on the surface.[51]

The following chapter turns to petroleum where these conflicts internal to the modern self become only more pronounced.

5 | A Faint Whiff of Gasoline

[She] had become ashamed of oil; and this was something new.
—Upton Sinclair, *Oil* (1927)

A wild sequence of traumas sits at the center of the most famous scene in petroleum's cultural history. That scene—the climax to George Steven's 1956 film *Giant*—depicts the Texas oil tycoon Jett Rink collapsing in a very public mental breakdown during a ceremony being held in his honor. With news cameras streaming the event nationwide and with the state's social elite gathered for the occasion, this Texas tycoon arrives at the ceremony sloppily drunk, weaving his way through the crowd; he sucker punches one of his guests in a confrontation over questions of honor; he retreats into a back room where he cowers pathetically in front of his main rival, who deems him not worth fighting; and he then finally and literally collapses to the ground in a heap of self-pity muttering to himself like a child: "Poor Jett . . . Poor Jett . . . Poor boy."[1]

Shame, violence, dishonor, and then finally infantile regression bind oil in this scene to human debasement and to a chain of traumatic events that plunge us somewhere into the troubled depths of the modern psyche. Oil and trauma—it is hard to imagine the metonymic associations being clearer than they were in this award-winning indictment of petroleum and its culture.[2]

What makes this scene from Steven's film a fitting opening to this final chapter on petroleum is that it reflects the broader moral condemnation of the new hydrocarbon economy that circulated in the United States during the rise of petroleum. Because Rink serves in the film as the personification of this alluring, yet disruptive new "petro-culture"[3]—with its volatile boom-and-bust strikes, its reckless economic behavior, its unfair political subsidies, and its toxic filth—he is set up to be the film's sacrificial lamb. His association with unearned wealth, conspicuous consumption, and gushing oil wells (or "bad oil," as Daniel Worden puts it)[4] does not sit comfortably with the film's pastoral

Jett Rink strikes oil: *Giant,* 1956. Archive Photos/MoviePix/Getty Images.

unconscious, which is yet steeped in a slower somatic world where cow-boys ride close to the earth, where men derive their wealth from the muscles of horse and cattle, and where westerners still stake their ener-gies and pleasures in the soil and grass that keep horse, cow, and rancher alive. Unlike the rest of us, Rink remains deaf, that is, to the film's loud lament against these unsettling changes, which petroleum augurs— "I won't have digging for oil. . . . Is that clear? This is a ranch, not an oil field"[5]—and thus his symbolic annihilation at the film's close is a sort of nervous commentary on the meaning of the nation's new oil depen-dencies and the bustling and anxious world oil had done so much to summon into life.

Oil and the Dialectic of Agency and Impotence

This Oscar-winning film reminds us that the symbolic import of oil, like that of coal and electricity, followed closely in step with its rising material significance in American life. By the time the film came out at midcen-tury, petroleum was both the basic building block of the nation's "mate-rial constitution" and a defining facet of people's embodied experience in their world.[6] To quote a contemporary of these changes, oil had satu-rated the nation's "culture complex" and the "faint whiff of gasoline" followed a person just about anywhere he or she went.[7]

Although oil's modern history dates back to the mid-nineteenth cen-tury, its cultural importance escalated sharply between the start of the twentieth century—when the internal combustion engine carried oil into what Stuart Chase called the "Power Age"—and midcentury when petroleum outpaced coal as the nation's predominant fuel source. Dur-ing this critical period, as we were transitioning to a people of prehistoric carbon, oil as a cultural object emerged as one of the most meaningful markers of our modernity—a sort of symbolic rallying point for what critic Peter Hitchcock calls our collective national efforts to represent ourselves to ourselves in this modern industrial phase of our history.[8] That is, as Americans grappled with the implications of living in a mod-ernized world of tankers, pipelines, wells, and the multiplying industrial and financial structures that they fueled, they saw modernity's intense antinomies in oil, including, on the one hand, the very real material and psychic joys that came with living in a fossil-fueled world and, on the other, the equally intense anguish that all too frequently accompanied the storm and stress of an oil-wrought modernization. Liberty and trauma, power and impotence, opportunity and injury—oil spoke to the tensions within a people torn psychically, and sometimes violently, between their eager embrace of modernity's new material freedoms and their underlying concerns about what it meant to leave behind a seemingly more manageable, and even human-centered, somatic world that was rapidly fading into history.

On the one hand, oil's material impacts afforded Americans a very real set of personal freedoms on which contemporaries placed a high value. With its rare energy density, its liquidity, and the portability of its prime mover, oil revolutionized the ways in which Americans both circulated and labored in their world. Because petroleum (when combined with the combustion engine) condensed the physical strength of hundreds of men, women, and horses into a single person's hands— without the unwieldy infrastructure required by the coal-fired railroads or the costly ecological infrastructure of slave quarters, stables, food, and fields required under the somatic regime—its energies translated in a concrete sense into a type of personal liberation from certain types of hard labor, from once impassable geographic barriers, and from the slow pace of foot traffic and muscle-powered work that had previously defined life. Observers consequently tallied among oil's emancipating impacts a dizzying array of social and cultural developments that included the modernization of farming, the migration to the suburbs, the Florida real-estate boom, the proliferation of golf courses, the roadhouse, the roadside hot-dog stand, and even such intimate matters as the shift in teenage dating rituals from front porch to backseat. The way social critic Lewis Mumford put it was that oil had gifted Americans a "power and mobility" that just a couple of decades before had not only been impossible but also almost unthinkable.[9]

For these obvious reasons, oil as a signifier fell in line with what was a ready-made narrative of technological progress that traced itself back to the Enlightenment, a muscular discourse of emancipation that defined release of the earth's physical energies as the material predicate for mankind's unfolding liberation from the natural world. The oil industry, for its part, capitalized on that established discourse by deeming the new material freedoms that came with oil a so-called petroleum bill of rights.[10] But few Americans in possession of a tractor or automobile needed corporate cheerleading to convince them that oil had liberated them from the limitations of the past. They could feel that power and freedom in the new ways their bodies moved through and labored in the world.

Yet oil's symbolic value always ran equally hard in the opposite direction. Despite its emancipatory potential, or perhaps even because of it, oil became a volatile flashpoint for broader public debates over the meaning of power and freedom under the terms of the nation's new petroculture. Although no one doubted that oil had liberated Americans in some perfectly literal ways, it was equally clear to contemporary observers that oil had also helped to coax into existence a highly structured world economy characterized by giant oil fields, tankers, pipelines, and financial monopolies that generated both new industrial-size barriers to individual opportunity and alarming ecological changes that were smothering out what remained of life under the somatic regime. In other words, not only did contemporaries worry over the sight and smell of nasty oil spills and monolithic oil fields but they also voiced a loud collective concern over oil's increasingly depersonalized economies of scale and giant monopolies that seemed to be undermining, or even cancelling out, the liberating potential of the nation's transition to fossil fuels.

The troubling dialectic of agency and structure, of power and impotence, that defined the public discourse of petroleum haunted oil for most of its earlier history, and although it can be found in many different places, it played out explicitly in a long-running cultural battle that was fought between oil's progressive critics and its boosters during the early and middle decades of the twentieth century when oil was rising in material importance.[11] In that culture war, oil's critics cast petroleum as a symbol of the structural pains and traumas of modern life under a privately controlled fossil-fueled regime, while its boosters—backed by well-financed public relations departments—suggested that oil and its culture were naturally in line with the nation's liberal-individualist values and pastoral traditions that stretched back in time to an earlier age of muscle power and foot traffic before fossil fuels had taken hold of national life.

The cultural imprint of that conflict was long lasting. On the one hand, it nurtured a polemic attachment to oil symbolized today by the puerile phrase "drill, baby, drill," and, on the other, it encouraged an equally blind rejection of the nation's oil dependencies on the left and right that has brought us no closer to appreciating, as Lemenager has pointed out, the depths of our embeddedness in a fossil-fueled regime, or moreover, how we might imagine our way out of it.[12]

Ashamed of Oil

We have always, to some extent, been "ashamed of oil," as one character from an early novel was.[13] The high-profile exposure of the Standard Oil monopoly in 1911, the US military's subversions of Mexican sovereignty to protect the nation's oil interests in 1917, the startling revelations of corruption over federal oil-lease sales in the Teapot Dome Scandal in 1924, the industry's poor conservation practices in the 1920s and 1930s, the alarming deployment of the Texas National Guard into the chaotic oil fields in 1931, and the persistent complaints over the unfairness of the oil-depletion allowance during the Eisenhower and Kennedy administrations—all meant that oil persisted as a beleaguered commodity in the national imagination.[14] And while oil and its industry might not have always deserved the nasty reputation they acquired in the twentieth century, they were nearly always painted in popular culture, as Roger and Diana Olien have shown, with a dirty brush.[15]

What that meant for oil as a signifier is that it circulated in the national imaginary as a sign of the many structural traumas of modern life. In the period's fiction and nonfiction alike, oil as a symbolic object evoked the high costs of modernization and the manifold risks that came with private ownership and control of the nation's limited energy resources. This deep-seated connection between oil and various types of socioeconomic, environmental, and psychic trauma ran deep. It was evident from the first reports that came out of the Titusville oil fields in 1859, which described them in terms of "disorder, disarray, and indifference,"[16] and it continued through morbid contemporary descriptions of the industry such as Paul Anderson's 2007 *There Will Be Blood,* which linked oil to human blood sacrifice. In such a context, Ida Tarbell's description of the nation's first oil tycoon, John D. Rockefeller, as a "living mummy," a kind of shrunken human artifact of the lifeless capital that structures and haunts modern behavior, is simply the most memorable metaphor for this pervasive anxiety that surrounded oil and the role it had played behind the scenes in American life.[17]

As an example of this discourse of oil and trauma in the period's nonfiction, we might turn to sociologist Robert Lynd's essays on the Standard Oil fields of Wyoming and Montana, where he spent six months as a roustabout and preacher. Those articles, "Done in Oil" and "Crude-Oil Religion," published in the middle-class magazines *Survey Graphic* and *Harper's* in 1922, speak to the same type of degradation and structural impotence that contemporaries had also equated with US coal towns.[18] According to Lynd, the nation's culture of oil, despite its obvious benefits, was being weighted down and compromised by the traumas experienced by oil workers and their families out in the oil fields and by the oil industry's embrace of an exploitative economic system that relegated modernity's traumas to the far reaches of the mineral frontier where gigantic parent companies such as Standard Oil of Indiana could keep them out of sight and legally distance themselves from the unsavory practices of the subsidiaries they funded.

"Oil town! Don't talk to me about oil towns. . . . I once lived in one of these oil towns. I know oil towns. Ugh! You never get away from the smell of the oil. You eat it and drink it and wear it and sleep it." Even before Lynd had arrived at the Standard Oil Elk Basin facilities, a passenger on the train had regaled him with sordid stories about the filth, squalor, and class exploitation he had in store for himself during his upcoming six-month stay in the northwestern oil fields. Out here on the far reaches of the mineral frontier, the passenger told him, one could expect to find a decidedly less civilized part of the country with none of the creature comforts, stability, or social progress implied by oil or by the material contributions it had made to a supposedly modernized nation. What Lynd found in the northwestern fields, in fact, replicated the same troubling patterns of class exploitation and arrested social development earlier writers had found in the nation's other marginalized geographies of energy production.[19]

As Lynd explained, the Wyoming and Montana facilities were hostile, primitive, and abject sites of production that clashed violently against the more celebrated geographies of modern life. These quasi-colonial spaces were, he explained, filthy, isolated, and mind-numbing environments in which to work and to raise children. The company town Lynd visited near Frannie, Wyoming (or to be perfectly clear in this case, the two-company town he visited) was not really even what one could call a town. "A handful—perhaps a hundred—of tents, tar-paper shacks and battleship-gray company bungalows," this makeshift encampment exemplified the type of underdeveloped communities that followed energy production wherever it went in the nineteenth- and early twentieth-century United States and abroad.[20]

The Elk Basin field was an especially atavistic place in that it was located in an isolated and rough part of the region's mountainous deserts and lacked most of the basic amenities one might expect from modern life. A literal "hole in the ground," according to Lynd—a mile wide, three miles long, 300 feet deep, and a mile above sea level—the basin was a "bleak," "monotonous," and "desolate" place that offered its workers and their families few natural comforts. There were, he said, "no water, no trees, no grass, not a living thing in sight, save the straggling sagebrush." A parched environment where water literally had to be hauled in from miles away by truck, this oil field offered itself up to Lynd as a ready metaphor for the poverty of workers being sucked dry by an uncompassionate industry.[21]

Moreover, the primitive conditions in which workers lived seemed to Lynd out of synch with the modern world. For example, housing (one of the four Malthusian essentials) was inadequate in the basin. Although some families lived in a limited number of company bungalows available to the field's more established workers, the bulk of the population found itself inhabiting canvas tents and shacks without the benefit of running water or any sort of formal plumbing. In these one-, two-, or three-bedroom hovels, families huddled together in "bare" little spaces that compacted all the functions of the house—"living-room-bedroom-everything-but-kitchen"—into a single room. Such poor housing was not simply a matter of niceties. It was a real concern in a region where seasonal temperatures vacillated wildly between "nauseous" heat waves in the summer, when the thermometer climbed above 100 degrees Fahrenheit, and subzero winds and blizzards in the winter, when infants and even healthy adults were at risk. In addition, the absence of any edifying cultural institutions, such as an established church, theater, union hall, proper school, or recreational center, contributed to the material poverty and stultifying isolation of the Standard Oil facilities.[22] In short, life was, Lynd said, "no joke" for oil workers and their families trying to have a go of it on the far reaches of the nation's mineral frontier.[23]

Other less life-threatening matters also made oil workers' material environment unhealthy and uncomfortable. As an example, hygiene in the camp was substandard, and workers had little control over this most intimate aspect of their own bodies. When Lynd visited the Elk Basin, there were only two communal showers for the basin's entire population of 500 people, and one of those showers had been out of commission during the previous winter when its pipes had frozen. Workers consequently went to bed sweatier and dirtier than they would have liked, and those who made the trek home "oil-soaked from head to foot," as Lynd described them, faced the unfortunate prospect of cleaning crude and paraffin from one's clothing and pores with only a bucket and wash-cloth in hand. Chronic bedbug infestations, the gasolining of floors and walls to get rid of those pests, and the pervasive smell of oil from the local production facilities meant that oil workers experienced a disquieting embodiment of petroleum that differentiated their laboring bodies from those of middle-class consumers empowered by their product many miles away.[24]

An equally harsh labor regime only reinforced the brutish nature of life in the Standard Oil facilities. According to Lynd, oil workers on the far reaches of the hydrocarbon economy, where federal oversight, public pressure, and union protections were nonexistent (a fact of life also found in the industry's Mexican and Saudi Arabian fields), found themselves at the mercy of sometimes uncaring employers who woke them up by horn each morning to the cursing of the local crumb boss, who required them to work a seven-day week without overtime pay, and who expected them to put in daily shifts of up to twelve hours—the so-called long shift—on a regular basis, thus making an already tough physical job a full two hours longer than that of their eastern counterparts. As Lynd explained, such naked practices of labor exploitation rendered middle-class platitudes about the virtue of hard work and mobility meaningless in a place like the oil fields. After a worker had finished twelve hours of "slogging away with pick and shovel at hardpan soil" or "rod-wrenching in slimy oil ooze," his body was, as one man put it, "good for nothing but John D. [Rockefeller]." Labor on the nation's oil frontier was, in other words, not the calling to God that the nation's middle class often imagined it to be so much as it was simply a working-class "hell."[25]

The point Lynd wanted his readers to take home from his portrait of the Standard Oil fields was that the traumatized bodies of workers and their families were no less a part of the nation's oil culture than were its teeming freeways, its newly mechanized farms, and its garish oil tycoons. "Gaunt" and "draggled" mothers, "barking" children, "stillborn" babies, and "bludgeoned" men were, he suggested, the hidden costs of the nation's growing dependency on a privatized and loosely regulated world of energy production that undermined the nation's more sanguine myths of progress. Yet to Lynd, these depressing conditions were ameliorable; they were not inherent to the production of oil so much as they were by-products of a ruthless industry that insisted on squeezing as much as it could out of workers' bodies while taking home returns on investments that reached as high as 4,500 percent.[26] In short, the type of damaged workers Lynd found laboring in the Elk Basin were simply, as he saw it, the human tinder for a private stratum of power that was out of synch with the very world it had helped to coax into life.

Lynd's portrait did not go uncontested. In fact, when word of Lynd's investigations reached the desk of John D. Rockefeller, his company's executives worked hard to silence the impact of those traumas on public opinion. The chairman of Rockefeller's Standard Oil of Indiana explained to Lynd (before publication of the article) that the company did not actually own these subsidiary operations and thus could not really be held accountable for their uncivilized practices (this despite the fact that Lynd had already learned that the Rockefeller interests possessed as much as 89 percent stock in the largest operation in the Elk Basin; that they controlled 93 to 97 percent of the total crude production in Wyoming; and that they dictated, according to the Federal Trade Commission, the pricing and conditions in the entire western, if not national, oil market).[27] Rockefeller also adopted a conciliatory approach by issuing a very public apology declaring that his subsidiaries' practices in the Elk Basin were "deeply concerning." That apology, however, came with a strategic repositioning of those traumas by claiming that they were in "marked contrast" to the industry's normal practices and unique to what Rockefeller called the "peculiarly barren, isolated, and difficult location" of the basin—this again despite the fact that contemporaries and subsequent historians have shown that work hours across the oil industry were exploitative and that a harsh racial, environmental, and class regime defined much of the industry's mineral frontier.[28] This concerted campaign to mitigate the bad press the oil industry received from Lynd's exposé came to its conclusion when a well-timed Rockefeller Institute grant redirected Lynd's attentions away from the dangerous subject of petroleum to the seemingly safer subject of Muncie, Indiana. At that point, Lynd's interests in the oil fields came to a close.[29]

Lynd's articles on the oil industry were, however, not unique. They reflected, in fact, a broader muckraking literature on the industry that characterized its practices as both uniquely predatory and out of synch with modern values. That ugly assessment of the industry, which cut across popular works such as Henry Lloyd Demerast's *Wealth versus Commonwealth,* Ida Tarbell's *History of Standard Oil,* John Ise's *United States Oil Policy,* and Matthew Josephson's *The Robber Barons,* focused explicitly on oil's contributions to the structural traumas of an overbearing industrial economy that seemed to be compromising the nation's liberal beliefs in a society of merit and mobility. The point of that literature was summed up perhaps best in Ise's claim that the oil industry and its investors were simply feeding off of a "gigantic system of wrong."[30]

The scope and intensity of that connection between oil and trauma is, however, hard to grasp without turning also to fictional works from the period that pressed home that point and that gave to it an emotional intensity and moral framework that left such a strong imprint on oil's subsequent life as a signifier. The breadth of this other fictional discourse of oil and trauma is vast. We can see it manifest in popular midcentury films such as *The Devil's Pipeline*, which associated oil with slave labor; *Beyond the Pecos*, which linked oil to physical violence; *Giant*, which tied oil to greed and psychic collapse; and *The Tiger Woman*, which connected oil to western imperialism.[31] Likewise, it is apparent in the period's poetry like Gary Snyder's "Oil," which turns to oil as a metaphor for "crazed" and addicted nations, Bertolt Brecht's "700 Intellectuals Pray to an Oil Tanker," which sees in oil an alienating American materialism, Pablo Neruda's "Standard Oil Co.," which casts oil as the symbol of economic rape, and Allen Ginsberg's "Who Owns America?" which associates oil with oligarchy and democratic decline. And it is, of course, apparent in more familiar contemporary works like *Syriana* and *There Will Be Blood,* in which the metonymic linkages among oil, violence, and trauma that have shaped our carbon culture for more than a century are decidedly direct.

Upton Sinclair's now-popular novel *Oil!* is, of course, the ur-example of this fictional literature on oil and trauma, both because it predates similar literature on the subject and because it distills so neatly into a single account the wide variety of economic, environmental, psychic, and social traumas we have come to associate with oil. Written in 1927 during California's oil boom shortly after the Teapot Dome Scandal, Sinclair's novel, which Lemenager terms an example of "peak-oil fiction,"[32] joined oil to a long metonymic chain of distress and decay that reflected the full spectrum of concerns contemporaries associated with oil (except perhaps the late twentieth-century link made between oil and foreign intrigue). Resource waste, toxic spills and fires, market manipulation, labor exploitation, pastoral decline, political corruption, and even suicide and murder—all found their way into this declensionist narrative of oil and its industry. While Sinclair could write joyfully about the somatic freedoms of driving in a car, as he did in the novel's opening scene, what he mostly saw in oil was the opportunity for a long and unhappy meditation on the nation's lost innocence in the wake of fossil fuels and the structures associated with them.[33]

Sinclair made this connection between oil and the structural traumas of modern life explicit in the novel's main plot and in several of its subplots. Myriad traumas pile up rapidly in the text in quick succession. They include the fleecing of local farmers by oil speculators trying to acquire mineral-rich properties through deception, the bribery of local politicians in order to shift the industry's production costs onto taxpayers, the secretive lobbying of federal officials to procure no-bid leases on public lands, the brutal exploitation of roughnecks in the oil fields by an antiunion industry, the devastation of nature by careless production practices, and the problems of building a stable social order out of a boom-and-bust resource economy characterized both by penury and meteoric accumulations of unearned wealth.

Although Sinclair's portrait of the nation's oil culture was drawn loosely from real-life events, including the Harding administration's recent Teapot Dome Scandal and the dramatic rise of western oil tycoon Edward Doheney,[34] Sinclair packaged those traumas in a simple parable concerning the industry's destruction of a rural town named (not too subtly) Paradise. This metaphor of a lost "paradise" in the wake of oil's discovery served in Sinclair's text to dramatize a popular collective lament for a simpler somatic age with its preindustrial rhythms and classical liberal values that preceded American entry into a crude oil–fired one that had all but displaced that premodern world and the people living in it.[35] In the text, the psychic, social, and material costs of integrating Paradise into the nation's advancing mineral frontier are played out figuratively in the deaths of a native son and daughter from that rural town. These twin deaths at the hands of the oil industry not only represent the novel's indictment of oil and its culture but also capture the intense and high-pitched emotional tone that linked oil to trauma in the national imaginary.

The first of those deaths concerns the relatively quiet suicide of a young country girl named Ruth. Described in the novel as a sweet girl of "honey and comb," a literal shepherdess who tends flocks of sheep like "boys and girls had done two thousand years ago in Palestine," Ruth circulates in the text as the personification of the American pastoral, standing in for a preindustrial world steeped in prelapsarian fantasies and Jeffersonian rhythms. Her psychological decline and eventual death in the novel symbolize for Sinclair his larger perception of oil's destructive impact on the nation's collective body. Ruth's suicide is, in effect, a strong symbolic statement of the human casualties effected by oil and its presumably pernicious version of modernity.[36]

Oil's destructive impact on Ruth is fairly direct. Her family's property first falls into the hands of the oil industry through a series of underhanded dealings. Next, her childhood home and pastures burn up in an oil fire caused by the industry's careless business practices. Shortly thereafter, her brother, a lay preacher, loses his moral compass as he is drawn into the nation's new advertising and consumer culture in the wake of oil discovery on their property. And in a final coup de grace that pushes Ruth over the edge of reason, her favorite sibling is literally beaten to death by oil industry thugs while struggling to build a union to give oil workers some control over their lives in the oil fields. For Ruth and thus for the pastoral culture that she represents, the promises of oil turn out to be a deadly bargain.[37]

The way in which Ruth ends her life is indicative of the psychological and physical violence contemporaries associated with oil. Having lost her brother and her home to the industry and unable to recover the joy of her youth, Ruth deteriorates psychologically by the end of the novel as she wanders catatonically across her family's former pastures and calls out after a dead brother and sheep that are long gone. As Ruth retreats psychologically in this scene into a lost pastoral age, wishing away the harsh realities of her present, she mentally degenerates, literally and symbolically, until in a final act of desperation she tosses herself into the original oil well that started her family's troubles. In this final death scene, with oil rigs rising up everywhere, the novel concludes on the macabre image of oil workers pulling Ruth's limp body out of the well with a set of metallic prongs. A heavy-handed gesture that links oil to physical and psychic violence, this scene plays out the loss of our former innocence by asking us to bear witness to one of modernity's casualties.[38]

And yet Ruth's muted death at the end of the novel is really secondary in importance to the more boisterous death of her brother, Paul, at the hands of the oil industry. Paul's death signifies a second type of trauma in the novel that is, like Ruth's, also directly connected to oil and its industry but not nearly so nostalgic or backward-looking in nature.

If Ruth represents the death of the pastoral in the wake of oil discovery, this other character, Paul, represents the loss of a subaltern alternative to the highly stratified world of private power that oil helped to coax to life. Whereas Ruth retreats into the past, Paul embraces the changes wrought by oil and grapples with modernity on its own terms. No Luddite, Paul accepts the inevitability of the industrial order that oil augurs, although he envisions it being built as a workers' paradise. In his transformation from rural goatherder to international labor organizer, Paul embodies for Sinclair the dream of an alternative worker-led world in the oil fields that is equally as modern as the one envisioned by oil's architects, yet presumably more mutualistic, equitable, and empowering for ordinary Americans. But Paul's fate is also tragic in the novel as he too falls victim to the oil industry and its violent practices.

This other tragedy is played out at the end of the novel in a wild and climactic scene that includes hired oil thugs pouring "boiling coffee" on workers' babies. During that climax, Paul is "lynched" by the industry and his skull cracked open in an effort to suppress the growing labor movement that he is leading in the oil fields. As Paul is rushed to the hospital with his head hemorrhaging from the blows he has taken, the text degenerates empathetically into a dizzying stream-of-consciousness narration centered on the scattered train of thoughts rambling through his traumatized brain. Jumping from Paul's lynching at the Industrial Workers of the World (IWW) headquarters in Long Beach, to his deathbed in the hospital, to memories of Russian workers marching in Red Square, to loud broadcasts of jazz blaring over the local radio station, to joyous shouts from the country's oligarchs celebrating Calvin Coolidge's victory, and to, weirdly enough, the heart of the Congo where "naked savages dance to the music of the tom tom, their black bodies smeared with palm oil," Paul's delirium suggests that the coming of oil has brought with it a more raucous—if symbolically and socially silenced—trauma than that which is announced by the nostalgic and acquiescent death of Ruth.[39]

The violence of the oil industry in this scene hurls the reader into a second traumatic landscape, a sort of figurative abyss where the rational distinctions we moderns tend to make between civilization and barbarism break down. In this scene, the schizophrenic nature of modern life, with its repressed dialectic of celebration and trauma, progress and barbarism, reason and irrationality rises like an oil blowout to disrupt the symbolic order of the text and to undermine the overly simplistic equation of oil and progress circulated by the industry's boosters. In this second scene from Sinclair's text, we see, that is, the material and psychic violence in modernity's basement rear its ugly head for a moment.

As it does so, it implicates oil not only in the demise of the pastoral but also in the traumatic birth of a modern petro-self.

Industry's Visions of the Oil Sublime

Oil bespoke trauma. The historical literature—both fictional and non-fictional—makes that perfectly clear. But oil was never simply about trauma. For all of the depressing rhetoric surrounding oil and its culture, both past and present, oil as a signifier has pulled equally hard in the opposite direction. Even oil's fiercest critics commonly understood oil—especially its technologies such as the automobile—within a teleological discourse of enlightenment that framed energy and mechanization as the keys to universal wealth and emancipation. Oil, after all, drove the engines of industry, facilitated trade beyond the barriers of the somatic economy, intensified the efficiency of food production, moved Americans out of crowded and disease-ridden cities, and emancipated many workers from the hard labor that went with clearing fields, harvesting trees, and tunneling for ore. The question for oil's critics such as Sinclair, Tarbell, Lynd, and Ise simply concerned the policies that were to regulate the production and consumption of that energy for the good of people and the environment. Like Sinclair's character, few of oil's critics were, in the last analysis, inveterate Luddites.

Oil's boosters never worried about the structural traumas associated with oil. They assumed—at least in their public statements—that the tapping of oil led directly to increased human liberation and that, therefore, the nation's petroculture did not need any particular rules or coaxing to bring it into democratic expression. Although that narrative could be found coming up from the grassroots at times, it was most vigorously promoted from the top down by the major players of the oil industry and their institutional voice piece, the American Petroleum Institute (API; founded in 1919), which launched what today we would call a culture war to rescript the meaning of oil in national life to emphasize its benefits and to downplay its ill effects. By midcentury, a steady stream of industry films, cartoons, educational videos, advertisements, press releases, and in-house magazines had circulated a much more reassuring narrative of oil that put a cap, so to speak, on the human and environmental costs of the petroleum age. While this earlier corporate discourse of oil was not as sophisticated as that to come later in the century, it prefigured what Sinclair had called the industry's efforts to "manufacture culture wholesale" along its own preferred ideological lines and economic interests.[40]

The industry's boosters worked hard to tether oil to a simple parable of progress that touted oil's liberating potential and that reconciled the dislocations caused by our dependencies on oil, at least symbolically, to the logic, rhythm, and norms of the somatic era that petroleum and its culture were otherwise rapidly displacing. Oil's boosters worked hard, that is, to systematically suppress the environmental and social disruptions caused by the nation's advancing oil frontier while repositioning criticisms of the industry as distractions from the real story of the industry's role in expanding opportunity, wealth, and freedom.

A film such as Robert Flaherty's award-winning *Louisiana Story*—a semidocumentary film that won the British Academy of Film and Television Arts award for Best Documentary in 1949 as well as a Pulitzer Prize for its musical score—is a prime example of how this other corporate discourse of oil worked to reconcile petroleum to the somatic economy it was displacing and to tamp down any budding concerns Americans had about the costs of living in a fossil-fueled world. The real importance of this professional collaboration between Standard Oil of New Jersey and the nation's leading documentary filmmaker was that in telling this semifictionalized story of Louisiana's advancing oil frontier, the rapidly globalizing culture of oil was made to appear not only innocent of injury but also continuous with (rather than disruptive of) the nation's somatic past and rural traditions.[41]

The plot of *Louisiana Story* centered on the drilling of an oil well in the Petit Anse Bayou in Louisiana's extensive lowlands. It begins with a gentle-talking oil contractor arriving by motorboat to sign a lease with a local Cajun family, and it then follows a member of that family, a local Huck Finn–like boy named Alexander Napoleon Bonaparte, for most of its plot as he plays and fishes in the swamp while building a relationship with the company's oil drillers and their business. The film's plot, which is secondary to the stunning cinematography that Flaherty took of the region, comes to a close when the oil well is finally brought in after some complications with the unexpected help of the boy. This Cajun boy's complicity with the project and the benevolent characterization of his relationship with the oilmen carries out the film's main message, which is, according to the Standard Oil contract, to show the audience the "benefits of oil to mankind" and (as the proposed script made clear) to demonstrate that "oil companies are beneficently public-spirited . . . and their operations productive and innocuous."[42]

What is most notable about *Louisiana Story* for our purposes is the masterful way in which it works to evoke a sort of oil sublime, or even oil pastoral, wherein the oil rig and its industrial culture are happily integrated into an imagined prelapsarian garden of raccoons, cranes, alligators, cypress knees, and bending reeds and wherein organism and mechanism are at perfect peace with each other. The film offers, in this way, a more hopeful version of oil's role in the remaking of the nation's culture complex by, among other things, sublimating the social, psychic, and environmental traumas that accompanied the drilling of oil wherever it went.

Louisiana Story explicitly acknowledges some of the sensory, psychic, and ecological traumas critics associated with oil. Hissing boilers, clanking machines, unruly "swamp buggies," and even a rig blowout make something of a mess of the bayou's calm in the film. But the film also works hard to neutralize those dislocations and to imagine a way toward rebuilding, as one critic of the film puts it, the "imaginative bonds between ourselves, our technologies, and our land."[43] For example, at the start of the film, when the rowdy wake of an oil-fueled powerboat tosses Alexander violently off his skiff while fishing, the sweet smile of the boy offsets any fears we might have about the nature of this loud and unprecedented intrusion. Likewise, the invasive act of drilling itself, represented in the film by a cacophony of industrial sounds coming from inside the rig, is humanized by the equanimity of a warm and loving oil driller named Tom who serves as a surrogate father to Alexander as he familiarizes himself with the new industrial landscape and soundscapes that oil brings to the bayou. And the traumatic blowout of the oil well itself, which releases uncontrolled high-pressure gas for several days in a bleak and worrying scene near the end of the film, is neutralized in the film's denouement by a strategic cut to a newspaper headline that reassures the viewer that "the skill of the modern driller" has "completely controlled" the unexpected release and brought in the well successfully. Moreover, the film assuages any remaining doubts that the viewer might have about the compatibility of this industrial culture of oil and this vernacular Cajun landscape in its final image, which shows Alexander perched on the oil well's rigging with his pet raccoon looking out on the bayou's tranquility. In this final image, boy, raccoon, marsh, and industry blur together into an imagined oil pastoral.[44]

The film, however, also works to naturalize the intrusion of oil into this region on a second powerful level by drawing on the authenticating power of the region's local Cajun culture. While the Cajun Bonaparte family authorizes the oil project itself, lending a sort of local legitimacy to the project, the film also works to legitimate the drilling project in a more subtle way through a series of symbolic pairings that downplay the radical novelty of oil's impact on the bayou. Alexander's daily fishing routine is, for instance, juxtaposed to that of oilmen fishing for oil; Alexander's struggle with a ferocious alligator is juxtaposed with the natural risks—such as a blowout—that come with drilling an oil well; and Alexander's Cajun father, a simple, loving, if decidedly vernacular, man, is juxtaposed with Tom, Alexander's surrogate father, this modern oil driller who seems equally loving and equally at home in the swamp. There is, in effect, nothing very new about the culture of oil that appears in the film despite the very real and monumental transformations it augurs.[45]

But the most explicit use of local Cajun culture to authorize the region's shift to a modern petroculture can be seen in the role that Alexander plays in bringing in the well at the end of the film. The film implies (even if fancifully) that it is this young Cajun boy, rather than the oilmen, who manages to bring in the well after several foiled attempts. Although Alexander's complicity in the drilling is clear from the beginning, his direct involvement comes at the end of the film when he takes it on himself to climb onto the oil rig and sprinkle some magic salt from his lucky fishing pouch and to pull out his most powerful talisman, a frog, to see what a little Cajun magic can do for this modernist enterprise. When the oil well comes in the next day, it is left to the imagination whether Alexander's fiddling with the project has had any effect, but the sequence of events merges the region's folk culture with the success of this industrial venture. Here the global mineral frontier and the vernacular landscape, modernity and tradition, oil and nature are safely at peace with each other as oil and its business are absorbed into this pastoral fantasy.

A film such as *Louisiana Story* suggested that oil and its culture were not at odds with the rhythms and values of the somatic world; it also suggested that the nation's advancing mineral frontier and its industrialized culture could be peacefully at home with the vernacular cultures and landscapes being transformed by those changes. That message was reinforced in other industry films such as *The American Frontier* (1953),

The oil pastoral: *Louisiana Story,* 1948. DVD capture. Home Vision Studio.

which did the same thing for the Norwegian cultures and the oilscapes of the Midwest as Flaherty's did for the southern swamps. But oil's boosters also worked hard to suggest that oil and its culture were not only compatible with the nation's rural traditions and local cultures but also in line with the nation's liberal traditions and values. The industry suggested in various ways that oil was, in fact, comfortably American and that its birth and its benefits had evolved naturally out of the nation's earlier liberal-individualist traditions and free-market commitments.

Industry and the Propagation of Oil's Origin Myth

We can turn to an API film such as the 1954 *Born in Freedom,* starring Vincent Price, to see how the industry also linked oil to an empowering rhetoric of expanding American freedoms rooted in liberal assumptions about the power and primacy of the individual. The main message of that film—that the "faith and determination of one man" alone brought us into the modern world—not only enshrouded the birth of oil in a comforting origin myth that tied oil and its culture back to long-held liberal fantasies about the freedom and autonomy of the individual but also guided viewers away from the concerns of oil's critics, who worried over both the structural constraints of the high-energy society and the growing influence in national life and politics of what they called the "Power Trust."[46]

Based on the life of the first man to strike oil, this thirty-minute historical fiction took the viewer back to the origins of oil, to mid-nineteenth-century Pennsylvania just before a little-known businessman named Edwin Drake struck the world's first oil well. The plot begins in 1858 when Drake (who called himself "Colonel Drake") is summoned by a New York banker to survey the prospects for collecting what he calls a "strange" substance seeping from one of the bank's properties up in the mountains. Appointed president of the bank's Seneca Rock Oil Company at the meeting, Drake subsequently packs up his bags and moves to western Pennsylvania to serve as the bank's sole agent on its new mineral frontier. But if Drake's appointment is quick and easy, his job as a prospector proves to be anything but that. It turns out that the region's rock oil, although easy to find, is exceedingly difficult to collect in viable quantities and that the region's locals, who think he is "crazy" for attempting such a scheme, want to "run [him] out of town" for his foolishness. Down on his luck out in the sticks, Drake's fortunes plunge even further when his investors also abandon him after several failed efforts to capture the oil by channeling it off the surface into holding tanks and then by digging holes to find its source. But throughout his trials, Drake persists with a dogged determination. A self-reliant man, Drake brings in the world's first oil well at the end of the film, and the narrator tells us that his success is the quintessential "American" story about freedom and independence and about how "one man's initiative" can change the course of history.[47]

Although *Born in Freedom* was neither a very good film nor a very successful one (even for Vincent Price), it demonstrates how this burgeoning corporate discourse of oil worked both to bring the culture of oil in line with the nation's liberal-individualist norms and to downplay the social and environmental costs of the nation's transition to oil. Two aspects of the film make that ideological link perfectly explicit.

The Story of Colonel Drake

A recently-released Technicolor movie tells the exciting story of Edwin L. Drake who overcame great odds to drill the world's first oil well at Titusville, Pa., in 1859.

Colonel Drake risked everything—and the oil industry was born. The story of Drake and his determination to drill for oil in the face of many handicaps typifies the entire industry today. Every oilman and every oil company must still take risks—despite all our improved techniques and scientific progress.

Because America's oilmen are free to take risks and free to compete for your business, the industry successfully meets the ever-increasing needs of the world's largest oil-consuming nation.

You benefit from the intense competition among thousands of companies—competition in the search for oil, in research, in refining, in transportation and at the service station. This rivalry brings you top-flight service, constantly-improved products, competitive prices.

That's how the oil industry, exemplified by Colonel Drake, works for you. It's a typical example of the way you benefit from America's system of competitive business.

Free! If you want to show this 16mm., 30 minute, color film "The Story of Colonel Drake" at your club, church or school, write, stating preferred showing dates, to: American Petroleum Institute, Box 178, 50 West 50th Street, New York 20, N. Y.

OIL FROM THE WORLD'S FIRST WELL! Climax of new movie shows Colonel Drake's faith rewarded as he happily examines first oil taken from the well he has just drilled at Titusville, Pa. Broadway premiere crowds found movie exciting because of many obstacles Drake (played by Vincent Price) overcame on the way to his historic discovery.

WHEN DRAKE came to Titusville, troughs were used to skim off oil coming from a spring. This impractical project was abandoned. Drake started from scratch.

FAILURE! Drake tried to dig a hole right down to the oil. Just as workers found traces of oil, water poured into the pit—nearly drowning the men. But Drake wouldn't give up.

JEERS and ridicule from the townspeople greeted Drake's attempts to find the source of the oil. After near-tragedy at the well, Drake had to face an angry mob. But he stayed on to try again.

"DRILLING" for oil was Drake's idea. With borrowed money he built a derrick. Again water flooded the hole. Failure was near.

"UNCLE BILLY," the driller, shared Drake's persistence. Though urged to quit they kept drilling and at 69 feet struck oil—opening new era for America.

TODAY'S oilmen, like Drake, face great odds. Despite today's scientific progress, only 1 out of 9 wells drilled, where oil has not been found before, ever produces oil.

YOU CAN MEASURE the oil industry's progress since Drake by today's powerful gasoline and motor oils, by the amount of service you get, and by the hundreds of other new products made from oil. This is how America's system of competitive business pays off for you.

Oil's origin myth: *Born in Freedom*, or *The Story of Colonel Drake*, 1954. Film Counselors.

First, like *Louisiana Story*, the film portrays the birth of oil—a remarkable, disruptive, thrilling, and paradigmatic event in the nation's history—as a perfectly familiar extension of the nation's history and traditions. The temporal framing of the film, which limits itself to the exciting lead-up to and discovery of oil, leaves out the subsequent—and arguably much more important—events unloosed by that discovery. Although we get a glimpse in a final photo montage of some of the physical changes that occurred in western Pennsylvania in the years to come, the film otherwise represses the well-documented social chaos, ecological degradation, and economic instability that went hand in hand with the discovery of oil in the region. The type of boom-and-bust real-estate speculation that sent people hurling into penury and riches, carcinogenic pollution that resulted from the uncontrolled dumping of oil into troughs and waterways, oil conflagrations that ravaged entire city streets, and the uprooting of established rural communities for the type of rough-and-tumble transient culture that characterized oil exploration were all left out of this myth although well-documented by contemporaries and known to us in the work of historians such as Brian Black.[48] The historically interesting stories of wild real-estate speculation and subsequent monopolization that followed on the heels of Drake's discovery were expunged from the record to allow for a more comforting mythology of oil.

Interestingly enough, the birth of oil, which remains innocent of its own history in the film, was also portrayed as a perfectly normal event in the nation's history rather than the decidedly novel one it was. After all, Petrolia, where oil was discovered, underwent an extreme makeover in the years to come after 1859, and the region's politics along with national economics experienced an irreversible change following the capture and monopolization of the nation's petroleum resources by a handful of major players in the industry.

But these events are downplayed in the film, and the discovery of oil appears instead to be a familiar event. The film's main supporting actor, a colloquial character named Uncle Billy, and his bright-eyed son work in the film to lend a certain backwoods familiarity to Drake's oil scheme and beliefs. Whereas local laborers fleece Drake for his money and middle-class onlookers jeer at him from the sidelines, this one rural artisan—who appears in the film in the dress of a simple frontiersman—alone believes in Drake and signs onto his oil scheme simply because, as he tells him, "when a man believes in himself as you do, it's kind of catching." A skilled salt driller by trade, this colloquial mechanic does more than simply facilitate Drake's drilling for oil. His vernacular speech, his rustic manners, his rural simplicity, and his artisanal independence all work to connect oil metonymically in the film to the type of frontier authenticity and republican simplicity that sits deep, even today, in the nation's political unconscious.

In fact, it is not even Drake who succeeds in bringing in the world's first oil well; it is Uncle Billy, who, while strolling along with his son on a fishing outing, stops by the well to pull out a lost drill bit. In the casual process of fishing the metal bit out of the shaft, it is Uncle Billy and his boy who first identify the telltale signs of crude oil dripping from the line. This original moment of discovery in the film works subtly to connect what was a radically new type of industrialized fishing for minerals beneath the earth to the much more traditional and homey ritual of father and son fishing for trout or bass on the weekend. What the film calls this "strange" substance (i.e., oil) turns out in the end to be almost familiar as it is midwifed by this vernacular artisan in a birthing scene that joins oil to a chain of signification that includes rural leisure, western nature, independent artisanship, and frontier families.[49]

If this birth of oil is decidedly more comforting than the traumatic birth of oil we see in critics' accounts, it is also framed as a story of American freedom rather than as a story of trauma, as is evident in the retitling of the film from *The Story of Colonel Drake* to *Born in Freedom* for mass circulation. This message of freedom is clear in the framing of the film. Starting with a long cinematic shot of a rural American main street with a conspicuous (and anachronistic) national flag hung from one of the buildings, the film rooted the viewer in the somatic unconscious of the nation's "horse-and-buggy age" even as it announced that the nation was about to leave behind this simpler preindustrial age because of Drake's singular resolve. This framing of the story as a patriotic story of individual initiative and national progress was latent throughout the film but was made most explicit in its ending, when the narrator boldly declares that Drake's victory over natural obstacles and collective doubt is a quintessential example of the power of "American ingenuity, freedom, [and] independence" and that, as we have already heard, "if one man's initiative and will" could accomplish such a feat, then certainly so could another's. After all, the narrator asks, "Who under these vast and free American skies would dare question the resolve of a man like Colonel Drake?"[50]

That stridently liberal and patriotic message about the power and primacy of the individual in the United States provides a dramatic closure to the film that links oil metonymically to the type of nineteenth-century values, traditions, and imagery that lay deep in the nation's political memory and self-identity and that originated in its somatic past. By returning the viewer to a preindustrial time before oil's structural impacts had been felt in any real measure, *Born in Freedom,* in other words, rooted this modern commodity in the nation's political unconscious by associating it with a classical discourse of freedom and empowerment that could be traced back to a western liberal-individualist tradition that took shape well before industrialization, corporatization, and the predominance of economies of scale took hold.

Moreover, the film's commercial advertisements, which suggested that Drake's story should be understood as a "typical example" of the way in which ordinary people benefited from the American "system of competitive enterprise," joined oil to a deeper national mythology that assumed the liberating potential of free markets in the absence of regulation.[51] To see how that emancipationist language of oil, freedom, and laissez-faire politics entered into the nation's oil ontology, we might turn from *Born in Freedom* to any number of other midcentury industry films, trade displays, advertisements, and educational shorts, each of which offered an iteration of this story of oil, free people, and free markets.

A revealing little piece of propaganda from 1956—an API cartoon entitled *Destination Earth*—provides an especially clear illustration of how this midcentury rhetoric of oil and freedom worked. That thirteen-minute animated short, distributed in schools, churches, theaters, and clubs, set itself up to be a playful ethnography of the nation's newly matured petroculture as seen through the eyes of an outsider. Its narrative linked oil explicitly not only to the material pleasures and freedoms petroleum had afforded Americans, such as those that came from driving an automobile or participating in the booming consumer economy, but also to the much broader set of political and economic freedoms that Americans presumably enjoyed. And while that wider narrative of oil and freedom was—and remains—a fairly suspect one, it had become a mainstay by midcentury in this other industrial discourse of oil.

Destination Earth was ideologically wrought in this regard. The plot centered on the exploits of a Martian named Colonel Cosmic who finds himself strapped to a spaceship for the sake of science and then hurled toward the planet earth (and more specifically, the United States) to study its dynamic free-market system and what the film terms its "fantastic" culture of highways, automobiles, and suburbs. Most of the cartoon simply follows the colonel around as he roams through the country's suburbs and cities jotting down notes on the planet's impressive gas-fueled automobiles, oil-heated homes, and various oil-derived goods, which have, as he observes, made life "better" for Americans than anyone else in the universe. The cartoon then wraps up back on Mars, where the colonel delivers his report to a coliseum of eager Martians waiting to learn the secrets of American success. In what turns out to be a rousing speech on the magical capacity of oil and free markets, the colonel presents to the crowd two totemic books, *The Story of Oil* and *The Story of Competition: More for Everybody*, that prove so inspiring that they cause a mass revolt on Mars in which the colonel's audience spontaneously throws off the yoke of political tyranny (the red planet is, of course, a totalitarian planet) and rushes out to drill for oil and build an entrepreneurial market society based on small-scale independent production. The final montage in this self-portrait of the industry finds a free Mars busily and happily wildcatting for oil while rapidly converting the planet into a retail-based consumer society.[52]

Although the cartoon itself is sappy, it helps us to see in bold relief the connection between oil and freedom that the industry was propagating in its broader cultural campaign in various other subtle ways. The most defensible claim it promoted in this regard was that oil had gifted Americans a set of new material freedoms, what Socony Mobil called in another context the "Petroleum Bill of Rights"—the "right to move around," the freedom from "human toil and sweat," the "free[dom] to live where [one] likes," the "free[dom] to look for a new job a hundred miles away," the "free[dom] to broaden [one's] understanding of their nation and the world through travel," and the freedom to trade beyond limited local "markets" with a larger and more dynamic global community.[53] *Destination Earth* dramatized those new material liberations by returning repeatedly to images of Americans' new automobility—to their rushing cars, teeming highways, and smiling faces that associated petroleum both with physical empowerment and a series of new psychic pleasures symbolized by speed, bustling people, and modern transportation technologies.

This petroleum bill of rights, however, implied more than the power of geographical mobility. It also implied *social* mobility and the expanded material opportunities that came with the transition to an

Oil, free people, and free markets: *Destination Earth,* 1956. DVD capture. American Petroleum Institute.

oil-based consumers' economy. "Opportunities for men, money, and materials" is the way General Motors put it at the 1939 World's Fair.[54] In this respect, glamorous portraits of middle-class affluence combined with a visual barrage of oil-derived consumer products such as toothbrushes, nylons, cosmetics, and synthetic footballs to link oil in this industry discourse to a narrative of social mobility that centered on the power of purchase. This connection, which can be found in industry propaganda at least as far back as the release of the claymation film *Pete Roleum and His Cousins* at the 1939 World's Fair, made oil into a signifier both of middle-class status and what historians have called, albeit mostly derisively, a "consumers' freedom" rooted in the freedom of choice over material goods.[55]

But the material freedoms that derived from the high-energy society only rarely appeared alone. More typically, they sat alongside and helped to mobilize a broader set of ideological claims about the relationships between oil and Americans' economic and political freedoms.

This broader claim about oil's liberating effects was a promiscuous one and in the case of *Destination Earth* took the shape of a cold war allegory. Whereas the colonel hails from an enslaved premodern world of autocrats, peons, and scarcity that chugs along on the basis of muscle power (or the Martian equivalent of muscle power), the United States appears as a free, fluid, and decidedly modern society that—as the result of modern energies—has managed to escape the age-old problems of scarcity and inequality that apparently still plague the rest of the galaxy. Whereas Martians run on hamster wheels, prodded along by a pinprick by their emperor in their buttocks, to generate energy for the red planet's power elite, we learn that the use of petroleum on earth has freed their American counterparts from a dependency on such literal energy slaves while serving as the material foundation for a free society where there is enough to go around for everyone. The implication is that oil lies at the bottom of what we might call this great interplanetary divergence, that it is the infrastructure of modern freedom, equality, and opportunity.[56]

At the same time, industry propaganda enjoined on this narrative of freedom a second rhetoric of free markets and their emancipatory qualities. In fact, this messy discourse of oil and freedom, which the API and the industry's major players circulated, could be boiled down to what was a simplistic and now familiar assertion about the liberating capacities of oil and laissez-faire policies. In *Destination Earth,* that message was articulated explicitly in the cartoon's climactic speech when we learn that the "big secret" behind American success is "not just a great source of energy [such as oil] but the freedom to make it work for everybody." As the colonel holds up to the crowd the two bibles of this secular society— *The Story of Oil* and *The Story of Competition: More for Everybody*—in a heavy-handed message linking freedom, equality, and prosperity to the oil industry and the magic of free markets, our Martian ethnographer ends on the rousing claim that "if you have both of these things . . . then any goal is possible!"[57]

This equation of oil and freedom offered a seductive alternative to critics' disquieting claims about the connections between oil and trauma and the need for deliberate regulations. Whether it appeared in visual texts such as *Destination Earth, Fire and the Wheel, 24 Hours of Progress,* and *Man and the Land* or in the industry literature that circulated in trade shows, magazines, and educational outreach, this discourse of petro-emancipation worked strategically to collapse into a simplistic and enduring rhetoric of oil and free markets the physical freedoms of the combustion engine, the material and psychic freedoms of the new consumer society, and the political freedoms of living in a democracy. Oil and freedom—that slippery chain of signification never quite went away.

To Manufacture Culture Wholesale

Few of us will doubt that oil was a liberating force in American life in some basic material ways, but its impact and history were always wrapped up in a decidedly more complicated dialectic of power and impotence. The industry's midcentury campaign to take control over oil's life as a signifier was, in effect, an organized effort to repress the uglier half of that dialectic. By burying the oligopolistic nature of the oil market beneath a rhetoric of free-market competition; by touting the virtues of free men and free enterprise without reference to the high entry costs to competition in a fossil-fueled age; by erasing the more complex story of state subsidies that propped up foreign and domestic investments in the industry (i.e., foreign wars, cheap leases on public lands, state-enforced rationing, and the oil-depletion allowance); and by shrinking, as we have seen, the meaning of freedom in public discourse to the freedom of consumption and the liberal rhetoric of laissez-faire econo-mies—this other narrative of oil and freedom put a cap on the traumas of the industrial age. Although this rhetoric of oil and freedom was yet in its infancy at midcentury—a far cry from the savvy public relations work that companies such as GE and Exxon undertake today—it re-minds us that defining the meaning of fossil fuels in the national imagi-nary was a political and ontological project that had implications both for what it meant to be American and what it meant to be dependent on cheap nonrenewable energy and the private industries that supplied it.

We can turn to a final industry film, appropriately named *A Fable*, for a fitting conclusion to the foregoing analysis of the public rhetoric of oil. This strange and revealing final text points us toward a certain fact about the future of energy discourse that bears telling.

In 1969, the International Division of the Mobil Oil Corporation made a peculiar decision to hire the pantomime Marcel Marceau to mime the virtues of free-market capitalism. The short, silent film that resulted from that collaboration centered on the message that personal freedom required free markets and that government regulation could only impinge on human happiness. What makes this film of interest to us, and what differentiates it from most of the other texts in this mid-century campaign, is that it effaced the role of oil in the script even as its message and very existence stemmed from oil's financial clout and the industry's self-interests.[58]

As the title suggests, the story itself took the shape of a fable. The film begins with a symbolic "everyman" (played by the mime Marceau) living alone in a pastoral fantasy, a sort of Garden of Eden, with fruit trees, clear skies, and rolling fields. In this solitary state, he is content, well fed, and surrounded by nature's untarnished beauty—a state of splendid isolation in which his solitude is only briefly interrupted for a brief rendezvous each morning with an attractive peasant girl who lives on the other side of a small stream. During that early morning interruption of their otherwise quiet lives, the two share a smile across the stream while they trade a couple of items, Marceau floating apples on a small handmade raft and this maiden, in turn, floating some homemade bread back to him.[59]

This state of peaceful isolation does not persist for long, however. Minutes into the film, Marceau is beset by unexpected visitors. The first intrusion comes when a troupe of children representing the different races and nations of the world arrive from nowhere. Marceau, delighted by the unexpected company, greets them cheerfully and plays and dances with them until he sees one errant child stoop down to pick a daisy from his field. In a sudden change of heart, Marceau banishes the children from his garden and builds a waist-high wall to keep these little interlopers out in the future. Marceau's troubles do not end there. Next Marceau finds himself beset by a group of equestrian hunters who jump over the wall to join him in a break from their pursuits. Again he welcomes these Turkish, English, Saudi Arabian, and Japanese riders into his garden until a second transgression occurs. This time one of the party's horses chews off a small branch of an orange tree. Again Marceau casts out the intruders and builds his wall higher so that neither children nor horses can clear the fence. The scene repeats itself until Marceau has built a wall and roof so secure that they keep out even the sun.[60]

Withering inside this prison Marceau has made for himself and lonely for companionship, this everyman has a change of heart. With the previous children's help, he dismantles the prison he has built around himself and then uses the materials to construct a bridge across the stream to get to the peasant girl. The film ends with Marceau and these happy children traipsing across the bridge to meet the girl on the other side of the stream while a message scrolls across the screen in six different languages: "If you lock out your friends and neighbors, sometimes your castle becomes your prison."[61]

The message here was that tariffs and regulations breed unhappiness. But no one who saw the film could have understood it to be an allegory for the dangers of economic nationalism. In fact, today the film circulates—if it circulates at all—as a children's story about friendship without reference to petroleum or energy policy. Still, the point of the film was that free markets breed happiness and freedom while regulated markets produce pale and joyless peoples.[62]

This promulgation of a covert rhetoric of freedom and laissez-faire politics, which we see played out in propaganda like *A Fable* and overtly in texts like *Destination Earth*, was standard fare in the industry's mid-century campaign to reshape both national and international popular opinion to embrace its own interests. What makes this particular film of interest, however, is that oil's presence was effaced from the transcript while oil's power was busily at work backstage producing that script. *A Fable* heralds, that is, a new and less transparent way of promoting oil's agenda. While the bizarre and "silent" strategy enacted in this film might have failed to get the industry's message across, it signaled the opening up of a broader cultural front that was less overt and both more creative and more insidious.

In the years to follow, the industry evolved only further from its rearguard defense of oil's reputation, which characterized much of its cultural output in the late 1930s, 1940s, and even 1950s, to this type of increasingly offensive and subterranean strategy, which used oil's economic weight to bend the public transcript on contemporary political and social issues toward its own interests. That advancing cultural front is familiar to us today, but it came to maturity as early as the 1970s when the soaring profits of Mobil Oil allowed it to launch a decade-long weekly editorial series on social problems opposite the *New York Times'* own op-ed page. With instructions to its ad men to keep the public off balance by taking "surprising" and unexpected positions for an oil company,[63] this more contemporary cultural campaign was the sign of a worrying world where it is exceedingly difficult to talk about drilling regulations, zoning laws, climate change, and the funding of renewable fuels without running headlong into the cultural clout of a private power that sits ever vigilant behind the scenes. That is simply to say that oil—like coal and electricity—continues to protect our addictions both by burying their social and ecological costs in the basement of the modern world and by removing our psychic stresses to the unarticulated depths of the modern self.

In the following conclusion, the costs of that repression come back to haunt us.

Conclusion
A Return of the Repressed

A few years ago, as part of its campaign to promote clean coal, General Electric (GE) aired a stunning advertisement. Dubbed "Model Miners," that advertisement treated viewers to an erotic montage of young models, both male and female, grinding their bodies down in a coal mine to the tune of Tennessee Ernie Ford's "Sixteen Tons"—a song about labor exploitation, hard fate, and working-class morbidity. To one journalist, it looked like GE had staged some sort of "stylish sex party" down in a coal mine.[1] But before being pulled from the screen (due to viewer complaints), that bold advertisement—which was more suitable to selling lingerie than the merits of coal and coal mining—pitted the ego against the id and put the sex drive on a collision course with reason. What, after all, was the brain to do when faced with half-naked young men and women grinding their bodies down in a coal mine to a song about working-class suffering on the mineral frontier?

Coal and sexual arousal—it was a long time coming, but somehow we got there.

As we have seen throughout this book, Americans have struggled to understand and to confront their dependencies on coal, oil, and natural gas. Because we have kept our energy dependencies mostly out of sight—by burying them geographically and socially in subaltern locations away from the nation's middle class, by repressing them symbolically within a bourgeois narrative of emancipation and growth, and by sublimating them psychically in the subterranean depths of the modern self—Americans have been more susceptible than we might otherwise have been to such wild misrepresentations of the world that we live in. Confronting those dependencies more explicitly and teasing out their ecological and ontological ramifications have been the unifying threads of this book.

Of course, we are still in the thick of modernity. Not much has changed over the past half century. There has only been a thickening in the web of our dependencies on prehistoric carbons, and today we only cling more tightly to them despite mounting evidence that the costs of doing so are rising and that we are on the verge of an existential threat to the ecological experiment we have been conducting for a century and a half. To bring this book to bear on the present, the following section thus offers five meditations on our continuing engagement with prehistoric carbon.

Reflection 1: A Return of the Repressed

First: We are witnessing a return of the repressed. The nation's energy crises are coming in quick succession now, and they have been since the 1970s. If Americans, including both the nation's middling and working classes, have had the privilege of ignoring their energy dependencies until recently (that is, since sometime after the state's completion of its massive electrification projects in the 1930s, the state's financing of a continental infrastructure of highways, and the democratization of the automobile), today it has become increasingly difficult to sustain that state of blissful neglect. The cartoonlike, and even infantile, calls to drill petroleum no matter what the cost and the vociferous denials of climate change are in historical context simply the desperate pleas of a portion of the public (reinforced by the industries and politicians that manipulate them) to insist the party is not winding down in the face of evidence to the contrary.

For the rest of us, the moral and material costs of combusting prehistoric carbon are becoming clearer. The underbelly of this deal we have had with the earth is popping out from underneath our clothing and is turning out to be as grotesque as we thought. One does not need to be a doomsayer nor even to countenance respectable theories such as peak oil to see that the price of energy has been rapidly rising since the last quarter of the past century and that it will continue to rise now that the rest of the world, particularly China and India, have signed onto modernity's ecology and emerged as major players in the world's carbon market. Our search for viable prehistoric carbons is consequently getting more frantic and our uninterrogated path dependencies are coaxing us into accepting new and higher risks and injuries.

Producing prehistoric carbon has always been dirty. But even so, our dependencies today are driving us into new, wildly destructive projects that are novel in their risk and scope. A case in point is Canada's Athabasca oil sands project, where there is a quantity of petroleum equivalent to 11 percent of the world's proven crude reserves (placing Alberta behind only Saudi Arabia and Venezuela in terms of proven reserves).[2] Getting that oil out, however, is ugly as it sits under 54,000 square miles of peat bog and arboreal forest. A resource-intensive process in its own right that requires large inputs of natural gas and fresh water and gives off high emissions, developing Alberta's sludgy bitumen requires either scraping off overburden, including trees, soil, moss, and clay, to strip-mine the desired carbon or heating up fresh water and injecting the steam it gives off deep into the earth to liquefy the crude and push it to the surface. Creating 400 million gallons of toxic water a day is a depressing prospect. The world's oil is, in other words, dirty and only getting dirtier.[3] Likewise, natural gas, a presumably clean carbon, is no longer simply the lucky by-product of petroleum drilling. US natural gas prospects, which could potentially meet the nation's natural gas needs for another century, nowadays rest on injecting a mixture of water, sand, and various chemical compounds into layers of shale with the hope of cracking those layers open to release energy-dense gasses. The current practice of pumping fresh water and chemicals into the planet's substructure, above which our aquifers rest, entails a set of risks that we are just now sorting out. Although it is not clear what these little human-induced earthquakes might mean in the long run, in the meantime they are fast becoming a part of our ecology and our lives.[4] And, finally, the continuing legacy of coal, a fuel source industry boosters assure us will sustain another 200 years of production, is backed by alarming strip-mining operations that have since the 1960s resulted in the removal of 450 mountaintops in Appalachia and the subsequent devastation of some of the nation's oldest and richest ecosystems. The largest carcinogenic spill in the United States (prior to the British Petroleum [BP] oil spill in the Gulf of Mexico) was, in fact, not an oil spill but an unpublicized 300-million-gallon coal-slurry pond collapse in Martin County, Kentucky, that sent a flood of coal's sludgy waste products into the sleepy mountain town of Inez in 2000.[5] Clean coal is, in other words, still very dirty coal.

Producing prehistoric carbons at the rates at which we consume them is the little problem. The big problem concerns combusting them under nearly any conditions. Whether we choose to burn modernity's fires in large-scale coal-generating plants or in the hundreds of millions of personal combustion engines, the ecological cost of our doing so is the emission of greenhouse gases such as methane, nitrous oxide, ozone, and carbon dioxide. Even natural gas, the cleanest of these options, releases methane and carbon dioxide along the path to its combustion. Hurricane Katrina, Hurricane Sandy, record freezes on the East Coast, and historically long droughts and high temperatures in the dry West signal that we are entering a new phase in our dependencies on prehistoric carbon and that modernity's ecology might be coming back to bite us. If the Intergovernmental Panel on Climate Change is even close to being correct, then we have burned our way into a new normal. In the United States, for instance, we should expect during the average year in the upcoming decades as much as 20 percent more rain and snow on the already wet East Coast, a drying up of the already dry West from as much as 30 percent less rainfall, and an increasing number of nonlinear weather events that pose an existential threat to water- and heat-sensitive crops, to coastal communities, to coral reefs and sea life, and to the more pedestrian experience of dipping into the ocean and feeling a summer breeze we once thought we understood.[6] Temperatures are expected to rise on the average by the end of the century to between 4 and 11 degrees Fahrenheit, and thus the unpredictability of our own 401Ks and Roth IRAs will be the least of our worries as we and our children are forced to adapt to this wildly novel ecology for which we have not evolved.[7]

Walking through the world will be different as we grow older in this century, and our combustion of prehistoric carbon will have had everything to do with it.

Reflection 2: Cultivating Ignorance

Second: We have strong incentives to continue to suppress the damage caused by our carbon dependencies. Those incentives can be traced back to two structural factors, one of which is at its root Machiavellian and the other of which is the product of human frailty.

Most obviously, the nation's private and public energy industries, at least those that specialize in prehistoric carbons, have very little incentive to change the rules of the game. A succession of energy crises has turned out to be very good for business. Seven of the top ten most profitable companies in 2012 were energy companies.[8] Exxon Mobil, despite a devastating oil spill in 1985, remains the nation's second-largest privately traded energy company, with earnings from oil and natural gas just shy of the world record in 2012;[9] BP, having recently weathered an oil spill the size of which threatened to shut down both the northern gulf fisheries and tourism industry and that many thought would shut down the company itself, instead made record pretax profits in the first quarter of 2013 and is beating all forecasts for recovery;[10] Peabody Energy, the world's largest privately traded coal producer, despite its industry's long history of bad press and growing competition from natural gas producers and rising production costs, still posted record revenues in 2012 and expects to dominate the global energy market in upcoming decades;[11] and abroad a once-sensible nation like Canada is now retooling itself as a wealthy petro-state amidst loud complaints about the Albertification of the nation. We are gearing up, in other words, for more of the same.

Those corporate earnings do cultural work, of course. They do not sit quietly on the sidelines. In fact, in the United States alone, approximately half a billion dollars were spent in 2008 to promote the nation's continued reliance on prehistoric carbons just as horse-trading and lobbying behind the scenes on K Street worked to ensure that the public debate over prehistoric carbons was framed in terms advantageous to the status quo.[12] In the 2012 presidential election, the fossil fuel industries had spent upwards of $158 million in television advertisements to promote their interests with still two months to go before the election.[13]

Today, it is difficult to imagine having a rational public conversation on the subject when the coal industry alone supports a vast cultural apparatus that includes such activities as funding public schools and sports events such as the Friends of Coal Bowl and its totemic trophy made of coal,[14] teaming up with popular media outlets such as the Spike cable channel to run reality-based television shows celebrating the thrills rather than human costs of hard-rock mining,[15] running front organizations that support coal with their friendly Facebook pages such as Friends of Coal, funding anti-climate-change research to run interference in the scientific consensus on global warming,[16] and using its financial muscle to discredit critics in such blatantly self-serving ways as to blame the high rates of infant birth defects in coal-mining regions on parents' supposed tendencies toward inbreeding.[17] If we look only at the case of coal, more than $40 million was spent in 2010 to convince the public that coal is clean (which it is not), that it provides lots of good jobs (which it doesn't), and that we might learn to relax and rely on it for another 200 years (which we can't unless 7 billion people figure out how to deal with rising seas, increasing droughts, razed ecosystems, and declining crop yields).[18] The point is that cumulatively the world's petrodollars, bituminous dollars, and natural gas dollars are playing an increasingly active role in muddying the nation's prospects for a balanced conversation over the use and regulation of prehistoric carbons. In such a context, we can understand GE's bizarre advertisement in 2008 as merely the uncomfortable cultural logic of where we might end up if this concerted work of repression goes on unchecked.

Yet manufacturing consent only works when we are inclined to let it work. The urge to look sideways at our energy dependencies goes well beyond unhampered propaganda and lax political contribution laws. It also derives from the fact that most Americans—on the political right and left and in the center—have very strong short-term incentives to want to believe that the status quo can be maintained. Our saturation in prehistoric carbon is so deep, so fully a part of the somatic and affective experiences of modern bodies, that we have evolved a built-in *will to believe* that is arguably as powerful as any corporate rationale. To imagine life without prehistoric carbons presumes a collective willingness to tear ourselves away from those somatic and affective attachments to the combustion of modernity's energies, it means divorcing ourselves from the thick memories of combusting carbon that are tied to our own personal and collective histories, and it means engaging ourselves in the very messy and uncomfortable work of finding out who we are and what we might be without combusting fossil fuels.

Disentangling ourselves from prehistoric carbon implies, in other words, that we are willing to cleave off a part of ourselves.

Reflection 3: Bodies by Carbon

Third: We have a persistent urge to look past the fact that modern bodies still recoil, contract, and suffer whenever prehistoric carbon is produced, transported, or combusted and that some of those bodies bear disproportionate burdens whenever modernity's fires are lit.

The human costs exacted by modernity's ecology are different than they once were, but those costs have not gone away. They have simply remained mostly out of sight. On the one hand, the globalizing of our energy markets has moved the mineral frontier to foreign and offshore locations where American bodies are less visibly at risk on a daily basis (unless we include in that tally foreign military interventions in the Middle East and the casualties that go with them). While domestic requirements for coal and natural gas can be mostly met within the geographic boundaries of the United States, the nation's oil portfolio, which makes up 35.3 percent of our energy consumption,[19] traces itself back to an unsettling portfolio of imports from eighty different nations that includes businesses operating in countries such as Angola, Nigeria, Chad, and Colombia, each of which has poor human rights records, weak environmental protections, and compromised political systems.[20] The so-called oil curse fuels political and human suffering in many places beyond the usual suspects. The case of Shell's activities in the Niger Delta, where state and industry have combined to destroy fragile wetlands and disempower poor coastal communities, is only a symbolic instance of our larger foreign-oil dependencies.[21] The dirty handshake, as one colleague has put it, that we accept each time we fuel up at the gas pump is, in a sense, the weekly ritual we enact to accept the many human and ecological externalities that are missing from the price we pay at the pump. These other costs are borne by other bodies and other ecosystems, and yet however far removed they might be from the American suburb and the national imaginary, those Nigerian bodies, Saudi Arabian bodies, and Colombian bodies are very much a part of us.

Likewise, subaltern bodies at home also continue to suffer whenever we produce and combust prehistoric carbon. But their injuries are less easy to document than they once were. If at one time nearly 750,000 coal miners went daily into the mines, putting their own bodies at risk and periodically making themselves heard in collective mass actions that gained public attention, that situation was changing dramatically by midcentury.[22] The automation of the mineral frontier since the 1940s, now defined by large strip-mining operations, has seen a plummeting in the unionized and nonunionized workforce in the nation's coal mines to about 90,000 workers.[23] Fewer bodies exposed to the sort of occupational hazards associated with underground mining's methane explosions and tragic mine collapses mean fewer bodies are collectively already organized to make a fuss about coal mining and its practices. Although unsafe dog holes continue to put some coal miners at risk, the structure of human injury in the United States is more indirect than direct.[24]

This collateral human damage of our carbon dependencies can sometimes be obvious. That damage is transparent anytime a new strip-mining operation in West Virginia forces poor communities to re-locate outside of a watershed, anytime a petroleum spill washes up on the shores of a coastal gulf community that lacks the financial resources to restore itself to normality, and anytime a slurry impoundment pond collapses and drowns out an already poor community in Kentucky. But more often the systematic damage borne by subaltern bodies tends to be less visible than that. It comes when coal-mining operations in Ap-palachia dirty up and reroute waterways, when they compel landscapes and soundscapes to undergo extreme makeovers, and when they push already disenfranchised communities into the sort of low-value extrac-tive resource economy that brings only a temporary salve to suffering. It is also felt when our poor, black, white, and Hispanic communities dis-proportionately find their children living next to highways and factories where the combustion of carbon leaves behind damaged infant bodies with high carcinogenic counts. And it is increasingly being experienced in the shape of extreme weather events, rising oceans, and the threat of reduced crop yields that will leave modernity's less-privileged bodies in a less resilient position to weather the proverbial storm.

If the world is on its way to being hot, flat, and crowded as a conse-quence of our practices, then we can expect it is going to be decidedly hotter, flatter, and more crowded for some bodies than for others.

Reflection 4: Stillborn Alternatives

Fourth: We are learning that the original Plan B might be a stillborn one. Atomic power, which promises to solve the short-term problem of reducing carbon-based emissions within a matter of decades, has arrived mostly stillborn in the United States as a result of the existential threat that it poses to life. Thus far, we have counted that risk equal in quantity if not kind to continuing to combust carbon.

That was not always thought to be true. At one point, deep in the fog of modernity, we were pretty sure we were going to wean ourselves off of prehistoric carbon without making sacrifices. We had, in fact, settled on a time and date: December 20, 1951, at 1:50 p.m. On that day, American physicists brought the big energy in uranium to bear on the more nu-anced civilian tasks of lighting up a room and burning a piece of toast.[25] If the undisciplined power released in radioactive chain reactions had previously been used to blow up parts of the planet, including cities such as Nagasaki and Hiroshima, small islands such as Bikini, and barren deserts such as Alamogordo, the nation's best and brightest minds had by midcentury trained it to do the more delicate tasks of lighting up four incandescent light bulbs in a sleepy little town called Arco.

Today, that historic facility, EBR-1, sits at the end of a dusty road in a lazy valley in Idaho, having fallen unhappily short of its promise. Although a few nations, most notably France, have managed since that time to successfully foist their energy burdens on uranium; several others, such as Japan and the Ukraine, have had some second thoughts about the consequences of doing so. The nuclear meltdown of the Fukushima facility in 2011, which followed a perfect storm of environmental events and human error in a rich industrial nation, and the Chernobyl meltdown in 1986 in a poorer one are simply the most recent reminders that nuclear fission carries a set of existential risks that arguably rate alongside the risks of carbon-induced climate change. It will be a bit longer before we know the full verdict on the long-term human, ecological, and financial costs of Japan's ill-fated rendezvous with nuclear power, but in the meantime the world's scientists and engineers are trying to work out the idea of a mile-long permanent wall of artificial ice that will keep the 300 tons of radioactive water (with its cesium, strontium, and tritium) currently coming from the plant out of the region's ocean ecosystem.[26] Whether or not that strange grappling with nuclear power's waste pans out, the dangers of Plan B will likely not be quickly forgotten.

Atomic power now supplies 5.7 percent of the world's energy requirements and about 9.1 percent of those of the United States. That number is not insignificant, and it will likely rise in the ensuing decades as the costs of combusting carbons accumulate.[27] The nuclear option, of course, presents a short-term fix to modernity's broken macroecology that would, assuming uranium supplies remain economically viable, lower the rate at which greenhouse gases are being emitted and thus slow down the rate of melting icebergs and bleached coral reefs. But the concern is that nuclear power risks rendering both large regional and aquatic ecosystems and the microecology of the human body hostile to themselves.

Changing modernity's ecology is a moral imperative. But the uncertainties trailing the nuclear option still persist. Leslie White put it this way in 1949: "The mastery of terrestrial fire was tolerable, but to create energy . . . is to play with celestial fire. Whether it can be done with impunity remains to be seen. The new Prometheus may also be the Executioner."[28]

Reflection 5: Utopian Alternatives

Fifth: We remain clouded in some large misunderstandings about the promises of going organic. Although it might be possible to go green nationally and globally with the right policy decisions and financial incentives, no one has produced clear and convincing numbers about how many people might be supported either at home or globally without some dependency on coal, oil, and natural gas or a ready substitute such as nuclear power. Rendering those numbers demands careful location-based accounting of soil fertility, water resources, and climate conditions that elude monolithic explanations, and it means reckoning with how greener options might supplant our current carbon-intensive global transportation infrastructure, which moves the world's goods around on oil-fueled railways, cargo planes, and long haul trucks. Moreover, how that transition might happen politically and culturally has not yet been worked out.

What we do know is that the sort of wild expectations placed on green biofuels such as corn ethanol, which reached a fever pitch early in this new century in the United States, were rooted in a fundamental misunderstanding of the world we live in. Biofuels were the rage for a few years, and in 2007, government mandates jumped in to support them[29] on the political logic that they represented a patriotic and green path for taking American fuel burdens out of the hands of foreign nations and shifting them back to the heartland, to Kansas and Iowa, where we could presumably grow our own fuel year after year without harm.

That promise of a green nationalism based on renewable fuels was an ill-fated one that quickly ran into what should have been some fairly obvious problems—the first of which was the problem of scale. Even under the most auspicious circumstances, converting our food farms into fuel plantations would only make up a fraction of the nation's and world's energy needs. The estimates are currently very rough, but a high-end estimate is that we might meet 15 percent of the world's energy needs by 2050 through biofuels, leaving us with large dependencies on non-soil-based fuels. Second, and more importantly, soil-based biofuels (imagined on a large scale) risk reversing modernity's gains by, in effect, placing our fuel dependencies back on the limited soil resources, that we had for so many centuries sought to overcome. According to one set of estimates by a World Bank research group on sustainability, we might need as much as 30 percent and 70 percent, respectively, of both North America's and the European Union's sugarcane, corn, wheat, sorghum, sugar beet, and cassava acreage to meet just 10 percent of the region's energy imports. Figures of that magnitude pose a problem because the world's arable lands are needed as farms to feed a rapidly rising population, as cotton and flax fields to clothe that population, and as forest reserves to provide the timber resources to house and supply a population that will likely hit 9 billion by 2050.[30] It is thus a risky business to demand that we recommission good farmland or even forest and pasture to take up our fuel burdens. And third, biofuels such as ethanol are tied up in already troubled commitments to industrial-farming technologies that are tough on the soil, tough on water resources, and hugely dependent on energy inputs in the first place. The ecological reverberations of the widespread use of fungicides and pesticides, the runoff of nitrogen into our waterways, and continued soil erosion are part and parcel of this supposedly green package, which would have biofuels scaled up for a crowded planet.[31] Biofuels might be part of the solution, but not yet, not exclusively, and not under these premises.

Our energy landscape is quickly evolving. Solar technologies, wind technologies, hydrogen cells, tidal energy farms, and geothermal energy capture all seem to be pointing in some meaningful directions—even if none presents a ready national or global solution. Yet within such an innovative and hopeful context, it is probably still worth being cautious in our assumptions in order to be clearheaded in our decisions. Riding a bike, recycling a coffee cup, putting in solar panels, and eating organic are symbolically central acts that presuppose important new existential positions and new ontological experiences, and they are acts that matter in some small ecological ways, but these little personal objections to modernity's ecology often come at the expense of digging more carefully and candidly into the much darker depths of the modern self.

That is, the task remains to understand why we continue to ooze carbon on the way to the local farmers' market.

Conclusion

In coming into our energy inheritance, we behaved badly—"like a drunken heir on a spree" was how Lewis Mumford put it.[32] Today, that spree seems to be coming to its climax. Somewhere along the way we took the road less travelled, and we don't have the option of returning to the point where the road first diverged. Way has led onto way, the modern self is in crisis, and so getting sober and waking up to where we have been during that spree might not be the worst place to start.

Appendix
Energy and Power

A few words are in order with regard to the two key terms in this book—energy and power.

In popular usage, the terms *energy* and *power* were, and are, used interchangeably, although to be sure, they are different things. Energy is the capacity to do work and is measured in, among other things, calories. Although it is maddeningly difficult to define exactly what energy is (Smil says we don't know if it comes in little "gobs" or what), we do know that it exists in fact and that it finds temporary residence both in material things, ranging from an apple on a tree to a lump of coal, and in such intangibles as the sun's rays, the weight of falling water, geothermal releases, wind, and electricity.[1]

This concept of energy, a founding concept in ecology and more broadly the natural sciences, allows us to tease out the complex ecological calculus by which the calories in coal, oil, and natural gas came to serve as the functional equivalencies of previous organic energy resources such as wheat, meat, trees, and grass. Thinking about energy in its abstracted form allows us, that is, to see how a modern culture, like the one we live in, has managed to radically expand its food supply, housing, clothing, and fuel sources—in other words, all of the Malthusian essentials relating to demographic and economic growth—by shifting its overall energy dependencies from the soil-based calories that could be derived annually from certain traditional energies like plant resources (i.e., food calories for human and animal prime movers and wood calories for heat, light, and shelter) to fossilized subsoil ones such as coal, oil, and natural gas that have taken pressures off of the soil and freed them up for other purposes. It helps us to see, that is, how fossil fuels permitted us to grow modernity's population and its economies under a different set of ecological terms than we could under what E. A. Wrigley calls the "organic economy," or what we would be able to do even today with the best practices in organic agriculture and forestry.[2]

It helps us see more clearly the threshold of modernity's ecology and why prehistoric carbons are so frighteningly important.

Power differs from energy in that it is the rate at which work is done rather than the capacity to do work. The term *horsepower* is our society's acknowledgment that the horse (which works at the rate of somewhere between six and twenty adult men) was the standard bearer of power, or work rate, for millennia before the coming of the steam and internal combustion engines. Thus when people speak, as Nikiforuk does, of modern Americans having the remarkable privilege of eighty-nine "energy slaves" at their disposal at the flick of a switch today, they are really imagining what it would be like to convert all of the energy that we use over the course of a day for our work, heat, light, and communication into actual physical work.[3]

The metaphor works nicely to capture the scope of our potential power, but it is in some respects an imprecise one. While the equivalency between energy and power is very direct when we climb into a two-ton pickup truck loaded with all of our belongings and drive it up a hill on the power of a 390-horsepower engine, the relationship is much more complex when we use that energy for purposes other than simply physical work. The trick to, and beauty of, modernity's ecology, and the difficulty in understanding it, is that coal, natural gas, and petroleum can today be easily converted into work, heat, *or* light, thus making fossil fuels the equivalent of a carriage ride to the theater, a warm home, a well-lit room, or (through a number of complex functional substitutions) a nice meal on the table.[4]

Notes

Introduction. Modernity's Basement

1. See Andrew Goodheart, "The Myth of the Machine," *Design Quarterly* 155 (Spring 1992): 24–28; David Nye, *American Technological Sublime* (Cambridge: Massachusetts Institute of Technology Press, 1994), 120–123; and Robert Rydell, *All the World's a Fair: Visions of Empire at American International Expositions, 1876–1916* (Chicago: University of Chicago Press, 1984), 15–17. For a contemporaneous account, see James Dabney McCabe, *The Illustrated History of the Centennial Exposition Held in Commemoration of the 100th Anniversary of the Declaration of Independence* (Philadelphia, PA: National Publishing, 1876), 294–295.

2. Phillip T. Sandhurst, et al. *The Great Centennial Exposition Critically Described and Illustrated* (Philadelphia: P. W. Ziegler, 1876), 368.

3. For a discussion of the theoretical framework that drove the planners' original designs for presenting raw materials, such as coal, see Rydell and Bruno Giberti, both of whom explain that planners hoped fairgoers would see how US raw materials (including coal) lay at the bottom of the nation's economic and cultural progress. But spatial constraints marginalized minerals on display, placing them primarily in the two mineral annexes behind Main Hall near the boiler room; and more generally, as Giberti has explained, the original discursive order of the fair fell apart for any number of reasons. See Rydell, *All the World's a Fair,* 20–21; and Giberti, *Designing the Centennial: A History of the 1876 International Exhibition in Philadelphia* (Lexington: University Press of Kentucky, 2002), 19–24, 219–225.

4. Goodheart, "Myth of the Machine," 28.

5. In the United States, coal surpassed wood as the nation's predominant fuel source in approximately 1885. Until 1951, it remained the nation's primary fuel source, supplying approximately 25 percent of US energy needs in 1870 and peaking at 72 percent in 1920. Thereafter, American reliance on petroleum and natural gas surpassed coal in terms of percentages, although coal, oil, and natural gas each continued to play an essential role in the nation's energy equation. See US Department of Energy, "Energy in the United States: A Brief History and Current Trends," xvii, http://www.eia.gov/totalenergy/data/annual/previous.cfm#1999 (accessed February 20, 2013).

6. See US Energy Information Administration, "Table 1.1: Estimated Primary Energy Consumption in the United States, Selected Years, 1635–1945," http://www.eia.gov/totalenergy/data/annual/showtext.cfm?t=ptb1601 and http://www.eia.gov/totalenergy/data/annual/showtext.cfm?t=ptb0101 (accessed February 20, 2013).

7. Stephen Kern, *The Culture of Time and Space, 1880–1920* (Cambridge, MA: Harvard University Press, 1983), 9. Kern discusses the material and psychic

excess—and the cultural adaptations—that came with the removal of previous organic energy constraints and our entrance into a world of cheap energy and ready power.

8. Peter Stallybrass and Allon White, *The Politics and Poetics of Transgression* (Ithaca, NY: Cornell University Press, 1986), 1–5.

9. J. R. McNeill, *Something New under the Sun: An Environmental History of the Twentieth-Century World* (New York: Norton, 2000), 10–16.

10. David Nye, *Consuming Power: A Social History of American Energies* (Cambridge: Massachusetts Institute of Technology Press, 1999), 187, 202, 206. For a history of electrification in the United States, see Nye, *Electrifying America: Social Meanings of a New Technology, 1880–1940* (Cambridge: Massachusetts Institute of Technology Press, 1992) and Ronald C. Tobey, *Technology as Freedom: The New Deal and the Electrical Modernization of the American Home* (Berkeley: University of California Press, 1996). For a discussion of the development of Americans' reliance on a national and global market of petroleum resources, see Robert Vitalis, *America's Kingdom: Mythmaking on the Saudi Oil Frontier* (Palo Alto, CA: Stanford University Press, 2007); the early chapters of Rachel Bronson, *Thicker Than Oil: America's Uneasy Partnership with Saudi Arabia* (Oxford, UK: Oxford University Press, 2006); and Chapters 5–6 of Richard H. K. Vietor, *Energy Policy in America since 1945* (Cambridge, UK: Cambridge University Press, 1984).

11. Andrew Nikiforuk, *The Energy of Slaves: Oil and the New Servitude* (Vancouver, BC: Greystone Books, 2012), 65. McNeill offers the estimate of seventy-five energy slaves in *Something New under the Sun*, 15–16.

12. See Chapter 6, "The Use of Energy," in Wendell Berry, *The Unsettling of America* (San Francisco: Sierra Club, 1996).

13. Vaclav Smil, *Energies* (Cambridge: Massachusetts Institute of Technology Press, 1999), x.

14. This sharp scholastic distinction that developed between our discussions of modernity's culture and its materiality was once the standard fare of such signature texts as T. J. Jackson Lears's *No Place of Grace*, Roland Marchand's *Advertising the American Dream*, and Louis Menand's Pulitzer Prize–winning *The Metaphysical Club*, each of which built its claims about historical causation around the terms *modern, modernity,* and/or *modernism* with little regard to the key material departures fossil fuels signaled in our species' relationship to the environment. See T. J. Jackson Lears, *No Place of Grace: Antimodernism and the Transformation of American Culture* (Chicago: Chicago University Press, 1994); Roland Marchand, *Advertising the American Dream: Making Way for Modernity* (Berkeley: University of California Press, 1986); Daniel Joseph Singal, *The War Within: From Victorian to Modernist Thought in the South, 1919–1945* (Chapel Hill: University of North Carolina Press, 1982); and Louis Menand, *The Metaphysical Club: A Story of Ideas in America* (New York: Farrar, Straus, and Giroux, 2002). See also Singal, "Modernism," in *A Companion to American Thought*, ed. Richard Wightman Fox and James T. Kloppenberg (Oxford, UK: Blackwell, 1998), 460–462.

15. That is not to say historians such as David Nye have ignored the environment in their work. The last chapter of Nye's *Consuming Power* and his later text

America as Second Creation both deal with how modernity's power technologies shaped and were shaped by the environment. My point (developed in the first chapter of this book) is that this focus on power is very different than a focus on energy and the bioenergetics of modern life. See Nye, *America as Second Creation: Technology and Narratives of New Beginnings* (Cambridge: Massachusetts Institute of Technology Press, 2003); Theodore L. Steinberg, "An Ecological Perspective on the Origins of Industrialization," *Environmental Review* 10, no. 4 (Winter 1986): 261.

16. Stephanie Lemenager, "Petro-Melancholia: The BP Blow-out and the Arts of Grief," *Qui Parle* (Spring/Summer 2011), 25–56; Thomas Andrews, *Killing for Coal: America's Deadliest Labor War* (Cambridge, MA: Harvard University Press, 2008); Anson Rabinbach, *The Human Motor: Energy, Fatigue, and the Origins of Modernity* (Berkeley: University of California Press, 1992); Nikiforuk, *The Energy of Slaves*; Kern, *Culture of Time and Space*; Richard White, *The Organic Machine: The Remaking of the Columbia River* (New York: Hill and Wang, 1992); Vitalis, *America's Kingdom*; Brian Black, *Petrolia: The Landscape of America's First Oil Boom* (Baltimore, MD: Johns Hopkins University Press, 2000); Myrna Santiago, *The Ecology of Oil: Environment, Labor, and the Mexican Revolution, 1900–1938* (Cambridge, UK: Cambridge University Press, 2006); Melanie McAlister, *Epic Encounters: Culture, Media, and U.S. Interests in the Middle East since 1945* (Berkeley: University of California Press, 2005); Carolyn de la Pena, *The Body Electric: How Strange Machines Built the Modern American* (New York: New York University Press, 2005); Imre Szeman, "Crude Aesthetics: The Politics of Oil Documentaries," *Journal of American Studies* 46 (2012): 423–439; Frederick Buell, "A Short History of Oil Cultures; or, the Marriage of Catastrophe and Exuberance," *Journal of American Studies* 46 (2012): 273–293; and, of course, the corpus of David Nye's work spanning three decades of publications on the interconnections among energy, culture, and western society, including *Consuming Power* and *Electrifying America*. See also the special issue "Oil in American History" in the *Journal of American History* 99, no. 1 (June 2012) and the other essays in the special issue "Oil Culture" in the *Journal of American Studies* 46, no. 2 (May 2012). On the pace and cultural adaptation to modernity's energy technologies, see also Joel Dinerstein, *Swinging the Machine: Modernity, Technology, and African American Culture between the World Wars* (Amherst: University of Massachusetts Press, 2003).

17. Leslie A. White, "Energy and the Evolution of Culture," *American Anthropologist* 45, no. 3 (September 1943): 335–356.

18. Elliott West, *Contested Plains: Indians, Goldseekers, and the Rush to Colorado* (Lawrence: University Press of Kansas, 1998), 332.

19. Michel Foucault, "Nietzsche, Genealogy, History," in *Language, Countermemory, Practice: Selected Essays and Interviews*, ed. and trans. Donald F. Bouchard (Ithaca, NY: Cornell University Press, 1977), 139, 146.

20. Ibid.

21. The concept of a global mineral frontier and its relationship to the American imaginary has been developed brilliantly in Vitalis, *America's Kingdom*, 18.

22. Sabrina Tavernise and Clifford Krauss, "Mine Owner Will Pay $209 Million in Blast That Killed 29 Workers," *New York Times*, December 7, 2011, A16;

Roger Alford, "Thousands of Kentucky Coal Miners Found Using Drugs," *Huffington Post,* February 23, 2012, http://www.huffingtonpost.com/2012/02/23/coal-miners-kentucky-thousands-using-drugs_n_1297173.html?ref=business (accessed February 22, 2012).

23. Cathy Caruth, *Unclaimed Experience: Trauma, Narrative, and History* (Baltimore, MD: Johns Hopkins University Press, 1996), 1–25.

24. The knowledge of trauma, according to critic Geoffrey H. Hartman, pulls in two different directions simultaneously: "One is the traumatic event, registered rather than experienced. It seems to have bypassed perception and consciousness, and falls directly into the psyche. The other is a kind of memory of the event, in the form of perpetual troping of it by the bypassed or severely split (dissociated) psyche." See Hartman, "On Traumatic Knowledge and Literary Studies," *New Literary History* 26, no. 3 (1995): 537–563.

25. Critic Amitav Ghosh discusses oil as being a matter of embarrassment that verges on the unspeakable and is nearly invisible and silent because of its immanence in our lives. See Ross Barrett and Daniel Worden, "Oil Culture: Guest Editors' Introduction," *Journal of American Studies* 46 (2012): 270.

26. Eric Reece, *Lost Mountain: A Year in the Vanishing Wilderness* (New York: Riverhead Books, 2006), 123–137.

27. According to Stallybrass and White, the symbolically low and the socially marginalized are intimately entwined in bourgeois consciousness and have historically followed a common psychic, spatial, symbolic, and social patterning. See Stallybrass and White, *Politics and Poetics of Transgression,* 1–5.

1. A People of Prehistoric Carbon

1. The phrase "a crisis of abundance" is from Stephen Kern, *The Culture of Time and Space, 1880–1920* (Cambridge, MA: Harvard University Press, 1983), 9. It refers to the material excess and psychic exuberance that attended the West's breaking through the energy barriers of the organic economy.

2. Theodore L. Steinberg, "An Ecological Perspective on the Origins of Industrialization," *Environmental Review* 10, no. 4 (Winter 1986): 273.

3. Peter Hitchcock, "Oil in an American Imaginary," *New Formations* 69 (Summer 2010): 81–97; Frederick Buell, "A Short History of Oil Cultures: Or, the Marriage of Catastrophe and Exuberance," *Journal of American Studies* 46 (2012): 273–293; Stephanie Lemenager, "Fossil Fuel: Manifesto for the Post-Oil Museum," *Journal of American Studies* 46 (2012): 375–394; Imre Szeman, "Crude Aesthetics: The Politics of Oil Documentaries," *Journal of American Studies* 46 (2012): 423–439; and Ross Barrett and Daniel Worden, "Oil Culture," *Journal of American Studies* 46 (2012): 269–272.

4. J. R. McNeill, *Something New under the Sun: An Environmental History of the Twentieth-Century World* (New York: Norton, 2000), 5–7; Rolf Peter Sieferle, *The Subterranean Forest: Energy Systems and the Industrial Revolution,* trans. Michael P. Osman (Cambridge, UK: White Horse Press, 2001), 78–137.

5. McNeill, *Something New under the Sun*, 8–10; Massimo Livi-Bacci, *A Concise History of World Population* (Malden, MA: Blackwell, 2007), 23–28.

6. Joseph K. Hart, "Power and Culture," *Survey Graphic*, March 1, 1924, 625; and Gifford Pinchot, "Giant Power," *Survey Graphic*, March 1, 1924, 521.

7. Vaclav Smil, *Energy in World History* (Cambridge: Massachusetts Institute of Technology Press, 1999), 93–102, 185–20; Anson Rabinbach, *The Human Motor: Energy: Fatigue, and the Origins of Modernity* (Berkeley: University of California Press, 1992), 45–68.

8. Kern, *Culture of Time and Space*, 10–36, 131–181.

9. Sherwood Anderson, "My Fire Burns," *Survey Graphic*, March 25, 1922, 997.

10. The terms *the great divergence* and *the great transition* have been developed most recently by Kenneth Pomeranz, *The Great Divergence: China, Europe, and the Making of the Modern World Economy* (Princeton, NJ: Princeton University Press, 2000), and Smil, *Energy in World History*. See also Edmund Burke, "The Big Story: Human History, Energy Regimes, and the Environment," in *The Environment and World History*, ed. Edmund Burke III and Kenneth Pomeranz (Berkeley: University of California Press, 2009), 33, 48; William Catton, *Overshoot: The Ecological Basis of Revolutionary Change* (Urbana: University of Illinois Press, 1980); Jean-Claude Debeir, Jean-Paul Deléage, and Daniel Hémery, *In the Servitude of Power: Energy and Civilization through the Ages*, trans. John Barzman (London: Zed Books, 1986); Sieferle, *Subterranean Forest*; and E. A. Wrigley, *Continuity, Chance, and Change: The Character of the Industrial Revolution in England* (New York: Cambridge University Press, 1988).

11. Although fossil fuels had been used, of course, in places such as the nation's great iron foundries of the East Coast a bit earlier in the nineteenth century, those developments were small and local. It was not until sometime around midcentury that the shift in the nation's energy base reached a critical turning point and the nation's railroads, steamships, and steel mills converted one by one from renewable wood fuel to nonrenewable coal, from nature's "free energy" to its "stored energy," from the organic economy to the fossil-fuel economy. See also Louis Hunter, *A History of Industrial Power in the United States*, vol. 2: *Steam Power* (Cambridge: Massachusetts Institute of Technology Press, 1991), 431. For national energy production and consumption statistics, see US Department of Commerce, Bureau of the Census, "Series M 76–92: Production and Calculated Consumption of Mineral Energy Fuels, Electricity from Waterpower, and Fuel Wood, in BTUs, 1800–1970," *Historical Statistics of the United States: Colonial Times to 1970* (White Plains, NY: Kraus International, 1989), 587–588.

12. These figures are offered by Chester C. Gilbert and Joseph Ezekiel Pogue, *Power: Its Significance and Needs*, Smithsonian Institution, US National Museum, Bulletin 102, pt. 5 (Washington, DC: Government Printing Office, 1918), 14. I take them to be relatively loose figures, given the methods of extrapolation.

13. Lewis Mumford, *Technics and Civilization* (New York: Harcourt and Brace, 1963), 156.

14. Gilbert and Rogue, 7, 14. The term *energy slave*, or in historical parlance *mechanical servant* or *electric slave*, has a long history, as seen in Chapter 2 of this

book, and it has been recirculated recently in texts such as McNeill's *Something New under the Sun* and Andrew Nikiforuk's *The Energy of Slaves: Oil and the New Servitude* (Vancouver, BC: Greystone Books, 2012). Moreover, Jean-Francois Mouhot has recently suggested that we might even want to begin to think about the moral equivalencies between our previous exploitation of actual slaves and our currently exploitative energy strategies, which are affecting the world's poor and nonwhite populations in disproportionately adverse ways. See Mouhot, "Past Connections and Present Similarities in Slave Ownership and Fossil Fuel Usage," *Climate Change* 105, nos. 1–2 (March 2011): 329–355.

15. Buell, "Short History of Oil Culture," 274.

16. Mumford, *Technics and Civilization*, 263.

17. For an assessment of the increases in real income and the material standards of living circa 1800 and 2000, see Gregory Clark, *A Farewell to Alms: A Brief Economic History of the World* (Princeton, NJ: Princeton University Press, 2007), 1–39, 194, 231. Clark, who is less interested in the ecological assumptions that go into the simple aggregate growth in the world gross domestic product (GDP), claims that what really divides the premodern from the modern world, ca. 1800, is the shift from essentially 0 percent productivity growth to about 1 percent annual productivity growth that amounts to a meaningful annual improvement for the first time in history in per capita material wealth. Of course, such figures fail to account for the continuing story of increasing economic inequality, wherein the gap between the rich and the poor nations of the world approaches 40:1. See also Smil, *Energy in World History*, 204.

18. The figures for population growth derive from McNeill, *Something New under the Sun*, 5–7, and Livi-Bacci, *Concise History of World Population*, 26–27.

19. Livi-Bacci, *Concise History of World Population*, 27.

20. Statistics from McNeill, "Table 1.1: Evolution of World GDP, 1500–1992," in *Something New under the Sun*, 6.

21. Figures derived from ibid.

22. Livi-Bacci, *Concise History of World Population*, 27; Smil, *Energy in World History*, 204; and McNeill, *Something New under the Sun*, 6.

23. Figures calculated from McNeill, "Table 1.1," in *Something New under the Sun*, 6.

24. Figures calculated from McNeill, "Table 1.3: World Population since 1820," in *Something New under the Sun*, 8.

25. Figures calculated from McNeill, "Table 1.1," in *Something New under the Sun*, 7.

26. Gregory Clark calculates the premodern world's growth rate in real income as more or less stationary. See Clark, *A Farewell to Alms: A Brief Economic History of the World* (Princeton, NJ: Princeton University Press, 2007), 193–194.

27. Lance Davis, *American Economic Growth: An Economist's History of the United States* (New York: Harper and Row, 1972), 21–22.

28. John J. McCusker and Russell R. Menard, *The Economy of British America, 1607–1789* (Chapel Hill: University of North Carolina Press, 1991), 53–60. See also Thomas Weiss, "US Labor Force Estimates and Economic Growth, 1800–1860," in *American Economic Growth and Standards of Living before the Civil War,*

ed. Robert E. Gallman and John Joseph Wallis (Chicago: University of Chicago Press, 1992), 35; Robert L. Heilbroner and Aaron Singer, *The Economic Transformation of America: 1600–Present* (New York: Harcourt, Brace, Jovanovich, 1984); Davis, *American Economic Growth*, 22.

29. Davis, *American Economic Growth*, 23, 123.

30. Ibid., 22.

31. Ibid., 22, 33–35.

32. Livi-Bacci, "Table 4.8: Population, GDP, and Productivity in 16 More-Developed Countries (1870 and 2000)," in *Concise History of World Population*, 130. See also Davis, *American Economic Growth*, who claims a 1.5 percent real annual growth rate between 1840 and 1960 (40).

33. Davis, *American Economic Growth*, 33–34. For growth rates in the 1960s, see US Department of Commerce, Bureau of the Census, "Series F 10–16: Growth Rates (Percent) of GNP and Output per Employee for the US and 6 Countries, 1870 to 1969," *Historical Statistics of the United States*, 225.

34. Davis, *American Economic Growth*, 124.

35. US Department of Commerce, Bureau of the Census, "Series A 6–8: Annual Population Estimates for the United States: 1790–1970," *Historical Statistics of the United States*, 8.

36. At the end of the twentieth century, when the US population reached 281 million, the United States had approximately 651 million acres of forest-use lands or 737 million acres total, 420 million acres of cropland, and 587 million acres of pasture. See Douglas W. MacCleery, *American Forests: A History of Resilience and Recovery* (Durham, NC: Forest History Society, 2011), 2, 14, http://www.fs.fed.us/pl/rpa/amforest.pdf (accessed August 21, 2013); World Bank, "Arable Land (Hectares per Person)," http://data.worldbank.org/indicator/AG.LND.ARBL.HA.PC (accessed February 22, 2013); Ruben N. Lubowski, Marlow Vesterby, and Shawn Bucholtz et al., "Major Uses of Land in the United States, 2002," US Department of Agriculture, *Economic Information Bulletin* no. EIB-14 (May 2006), http://www.ers.usda.gov/media/249896/eib14_report summary_1_.pdf (accessed August 21, 2013). To get an estimate of the 1860 per capita acreage I have simply assumed that the land in use today could have potentially been developed in 1860 for the same purpose when the population was 31 million. The numbers take into account Alaska and Hawaii, which were not yet part of the nation's land base.

37. Joachim Radkau offers an interesting and excellent counterargument to this notion of eighteenth-century crisis by claiming that "the limits of the 'solar energy system' had by no means been reached" and that the crisis was only a crisis if one assumed "an inexorable, continuous growth of population and commercial activity." In Radkau, *Nature and Power: A Global History of the Environment*, trans. Thomas Dunlap (Cambridge, UK: Cambridge University Press, 2008), 204.

38. See Thomas Robertson's counterclaim that Malthusianism runs through the American environmental movement's concerns in the twentieth century. Robertson, *The Malthusian Moment: Global Population Growth and the Birth of American Environmentalism* (New Brunswick, NJ: Rutgers University Press, 2012).

39. Quoted in Michael Watts, "A Tale of Two Gulfs: Life, Death, and Dispossession along Two Oil Frontiers," *American Quarterly* 64, no. 3 (September 2012): 439.

40. See Catton, *Overshoot.*

41. Burke, "Big Story," 33, 48.

42. The term *subterranean forest* comes from Sieferle's book by that name, *The Subterranean Forest.*

43. Howard T. Odum, *Environment, Power, and Society* (New York: Wiley, 1971), 115.

44. Gilbert and Pogue argue that we should assume one mechanical horsepower unit to be the equivalent of twenty units of manpower (rather than the standard six to ten) because of the limited endurance of humans relative to machines. This adjustment accounts for their much higher estimate than other writers such as Chase. See Gilbert and Pogue, *Power: Its Significance and Needs,* 14.

45. See Leslie A. White, "Energy and the Evolution of Culture," *American Anthropologist* 45, no. 3 (September 1943): 340, and *The Evolution of Culture: The Development of Civilization to the Fall of Rome* (New York: McGraw-Hill, 1959), 41.

46. Smil, *Energy in World History,* 40–49; Ann Greene, *Horses at Work: Harnessing Power in Industrial America* (Cambridge, MA: Harvard University Press, 2008).

47. Nikiforuk, *The Energy of Slaves,* 3.

48. Debeir et al., *In the Servitude of Power,* 28–33.

49. Kenneth F. Kiple and Virginia Himmelsteib King, *Another Dimension to the Black Diaspora: Diet, Disease, Racism* (Cambridge, UK: Cambridge University Press, 1981), 88.

50. Debeir et al., *In the Servitude of Power,* 4.

51. The way E. A. Wrigley puts it in his fastidious study of early modern England is that this somatic, or as he calls it "organic," energy regime was always caught up in a "negative feedback" loop in which getting more productivity out of the land required more labor, and increasing labor, in turn, required more pressures on that same land. The system was, he tells us, landlocked in a sense, stacked against itself simply because labor, shelter, fiber, and fuel all ultimately competed with each other and nudged each other out for space on the soil. Without additional labor power, economies did not expand; without more land to grow food and fuel, labor power was more or less static. The premodern pressure imposed on soil resources by growth is what ultimately pushed England to supplement its fuel and labor power with coal and to seek out soil resources elsewhere by taking colonies. Wrigley explains that in contrast to modern economies, the economies of the preindustrial world were thus defined by diminishing returns in which land constraints dictated the limits of economic and demographic growth. Classical economics, once referred to as the "dismal science," was founded on this debilitating fact, the theories of Thomas Malthus and those of Adam Smith merely two sides of the same coin. All premodern ("organic") societies were caught up in an economic cul de sac. As Wrigley puts it:

An organic economy, however advanced, was subject to negative feedback in the sense that the very process of growth set in train changes that made further growth additionally difficult because of the operation of declining marginal returns in production from the land. It was their appreciation of this point that induced classical economists to take a pessimistic view of future prospects, especially in relation to real wages. Each step taken made the next a little more painful to take. In parts of an organic economy, because of the effect of specialization of function, increasing returns were obtainable and positive feedback existed, but, since each round of expansion necessarily increased pressure on the land by raising demand for industrial raw materials, as well as food, in the system as a whole negative feedback tended to prevail.

The widespread adoption of coal and steam set in motion a different train of resource possibilities no longer tied to the cultivation and exploitation of arable land. Coal and steam not only took the place of firewood as fuel but also came to affect dramatically the production of the other Malthusian essentials like housing, food, and clothing. Wrigley, *Continuity, Chance, and Change*, 29.

52. Pomeranz, *Great Divergence*, 241.

53. Ibid., 211–285.

54. Stuart Chase, *Men and Machines* (New York: Macmillan, 1929), 74.

55. Estimates are that the United States was using 1.6 billion horsepower from all prime movers in 1930; only 17 million was derived from work animals, and the vast majority of that horsepower, 1.4 billion, was derived from automotive sources. In US Department of Commerce, Bureau of the Census, "Series S 1–14: Total Horsepower of All Prime Movers: 1849–1970," *Historical Statistics of the United States*, 818.

56. Ibid., "Series K 430–444: Indexes of Farm Output per Man-Hour: 1939–1970," 500. See also Livi-Bacci, *Concise History of World Population*, for worker productivity, 130.

57. Smil, *Energy in World History*, 73.

58. The nation had about 25 million head of horses and mules, 16.5 million of which were primarily used for draft power. Ibid., 91.

59. This rough estimate is based on the assumption that the United States has had in the twentieth century about 1 billion acres of pasture and cropland available to be allocated for food production, and the US population was hovering around 106 million in 1920. Not all of that land was in use during this period, but its existence conditioned assumptions of growth. Lubowski et al., "Major Uses of Land in the United States 2002." For a portrait of early modern soil pressures in Europe and China, see Pomeranz, *Great Divergence*, 223–230.

60. Chase, *Men and Machines*, 72. Chase's figures actually underestimate the usage of mechanical horsepower, because he was writing at the very moment automobiles (our biggest consumer of energy) were entering into the American middle-class energy budget. In 1920, the United States had available about 430 million mechanical horsepower (even if it was not all in use at any given time), but by 1930 that figure had climbed to approximately 1.6 billion mechanical horsepower units. To imagine coming up with the grazing lands (or horse feed)

to sustain over a billion and a half horses, each of which would have consumed when working about 4 kilograms (or 8.8 pounds) of grain a day or 1.4 metric tons of grain annually if working, theoretically, every day (when grain yields were 1.5 metric tons per hectare) and 2 kilograms of grain a day when not working means we would have needed more than half a hectare per draft animal, or as much as 2 billion acres of dedicated cropland for our draft animals. See Smil, *Energy in World History*, 91, for calculations on draft animal requirements. Smil uses early twentieth-century crop yields for his assumptions about yields per acre. For an estimate of the mechanical horsepower available in the United States by decade, see US Department of Commerce, Bureau of the Census, "Series S 1–14," 818.

61. Chase, 9.

62. These figures are meant to be illustrative, and they are rough. Consumption of energy for the purposes of power as opposed to its value as heat or light is difficult to parse out in an age when power and energy are so easily interchangeable. According to Chase, about half of US energy consumption went into producing power, and about half (he estimated it at 500 million tons of coal equivalent [TCE] ca. 1920) went to the purpose of heat production rather than work. Ibid., 25, 91. The figure is probably a bit less because contemporary estimates are that Americans were consuming 21.3 quadrillion British thermal units (BTU) in 1920 or about 831 million tons of coal equivalent total, a figure that, if halved, is lower than that which Chase provides. See "Series M-76–92: Production and Calculated Consumption of Mineral Energy Fuels, Electricity from Waterpower, and Fuel Wood, in BTUs: 1800–1970," 588.

63. Pomeranz and Smil estimate that the sustainable yield of one hectare of an early modern forest was 3.6 tons of wood (or 1.8 TCE) for every hectare, or 2.47 acres of forest, or .73 TCE for each acre. It would take, under those terms, about 686 million acres of sustainably harvested cordwood to meet the nation's heat budget if we accept Chase's estimate of 500 million tons of coal as an acceptable approximation. See Smil quoted in Pomeranz, *Great Divergence*, 308. It is possible, of course, to imagine a well-managed coppice with modern fertilizer inputs giving a much higher yield of ethanol, wood, or bamboo, but the early modern figures for forest production are more meaningful because those are the ones that we would have been dealing with under the earlier ecology of production. More recently, in 2013, Americans consumed approximately 95 quadrillion BTU, or about 3.5 billion TCE, which would equate under these terms to more than 4 billion acres of forests, far more than four times what we had in the United States at any given point in recorded history. See US Energy Information Administration annual report at http://www.eia.gov/energy_in_brief/article/major_energy_sources_and_users.cfm (accessed August 19, 2013).

64. William Cronon, *Changes in the Land* (New York: Hill and Wang, 1983), 120–121. For a more modest estimate of the use of cordwood in the late colonial New England economy, see Brian Donahue, *The Great Meadow: Farmers and the Land in Colonial Concord* (New Haven, CT: Yale University Press, 2007), 176–179.

65. McCusker and Menard, *Economy of British America*, 29, 108–110, 170–171, 174, 184, 198–199, 204–205, 314–321.

66. William Cronon, *Nature's Metropolis: Chicago and the Great West* (New York: Norton, 1992), 148–206.

67. Thomas Andrews, *Killing for Coal: America's Deadliest Labor War* (Cambridge, MA: Harvard University Press, 2008), 83–86.

68. Cronon, *Nature's Metropolis*, 151–206; Jack Temple Kirby, *Mockingbird Song: Ecological Landscapes of the South* (Chapel Hill: University of North Carolina Press, 2008), 135–136.

69. US Department of Commerce, Bureau of the Census, "Series A 288–319: Households, Families, Subfamilies, Married Couples, and Unrelated Individuals, 1790–1970," *Historical Abstracts of the United States*, 41.

70. The figure above is derived from the assumption that the minimum requirement for per capita consumption of fuel, as estimated by Pomeranz, is about .5 TCE based on his estimate for what the early modern Chinese and Europeans required for their fuel needs, or what would translate into about 53 million TCE for Americans' minimal needs in 1920 to have an adequate standard of living. An acre of forest could sustainably produce approximately .7 TCE, so that would have required about 75 million acres of forests set aside in perpetuity to meet the minimum requirements of Americans in the 1920s. That requirement would have been easily met. Pomeranz, *Great Divergence*, 308–309. For approximate historic forest coverage, see Sonya Oswalt, Mike Thompson, and W. Brad Smith, eds., *US Forest Resource Facts and Historical Trends*, Forest Inventory and Analysis National Program/USDA, 4, http://fia.fs.fed.us (accessed March 22, 2013).

71. US Energy Information Administration estimates are that in 1920 US energy consumption was 21 quadrillion BTU, or the equivalent of about 755 million TCE, or about 7.1 TCE per person. These calculations are based on the assumption that an early modern forest could yield sustainably 1.8 TCE annually per every 2.47 acres. If we push the story forward, by 2011, Americans were using approximately 312 million BTU per capita, or 11 TCE per person. Using the previous assumptions, Americans today would need about 15 acres per person to supply their fuel needs from an early modern forest. See US Energy Information Administration, "Quick Facts," http://www.eia.gov/tools/faqs/faq.cfm?id=85&t=1 (accessed February 22, 2013).

72. Oswalt et al., *US Forest Resource Facts and Historical Trends*, 4.

73. Catton, *Overshoot*, 45.

74. Sieferle, *Subterranean Forest*, 64–65.

75. US Geological Survey, "Iron and Steel Statistics, 1900–2009," December 7, 2010 (accessed February 22, 2013).

76. Debeir et al., *In the Servitude of Power*, 11.

77. Sieferle, *Subterranean Forest*, 61.

78. Ibid., 64.

79. Ibid., 137.

80. Jeffrey L. Meikle, *American Plastic: A Cultural History* (New Brunswick, NJ: Rutgers University Press, 1995), 138, 148–149, 213.

81. Luis Fernández-Galiano, *Fire and Memory: On Architecture and Energy* (Cambridge: Massachusetts Institute of Technology Press, 2000), 8–9.

82. Mumford, *Technics and Civilization*, 151.

83. Livi-Bacci, in *Concise History of World Population*, estimates that a unit of energy in 1978 produced three times the wealth it did in 1850, a startling

efficiency in energy use that is simply dwarfed by the growing material consumption per capita since that period (124).

84. Chase, *Men and Machines*, 222–223.

85. University of Michigan Center for Sustainable Systems, *U.S. Material Use Factsheet*, Publication Number CSS05–18 (2012).

86. Joachim Radkau, *Wood: A History*, trans. Patrick Camiller (Cambridge, UK: Polity, 2012).

87. McCusker and Menard, *Economy of British America*, 314–321. See also, for example, Fred Kniffen and Henry Glassie, "Building in Wood in the Eastern United States: A Time-Place Perspective," in *Material Culture Studies in America: An Anthology*, ed. Thomas Schlereth (Lanham, MD: Altamira Press, 2006), 237–250.

88. US Department of Commerce, Bureau of the Census, "Series 550–563: Hay, Cotton, Cottonseed, Shorn Wool, and Tobacco—Acreage Production, and Price: 1790–1970," *Historical Statistics of the United States* 517; US Department of Commerce, Bureau of the Census, "Series L 87–97: Per Capita Consumption of Timber Products, by Major Product: 1900–1970," *Historical Statistics of the United States*, 540. The figures are based on the assumption that the United States had about 408 million acres of cropland in the twentieth century and had about 106 million people in 1920.

89. Catton, *Overshoot*, 41.

90. Chase, *Men and Machines*, 108.

91. US Energy Information Administration, "How Much Oil Is Used to Make Plastic?" http://www.eia.gov/tools/faqs/faq.cfm?id=34&t=6 (accessed on February 27, 2013).

92. Janet Ore, "Mobile Home Syndrome, Engineered Woods, and the Making of a Domestic Ecology in the Post–World War II Era," *Technology and Culture* 52, no. 2 (April 2011): 267.

93. I. B. Holley, "Blacktop: How Asphalt Paving Came to the Urban Environment," *Technology and Culture* 44, no. 1 (October 2003): 703–733.

94. Quoted in Jeffrey Meikle, "Material Doubts: The Consequences of Plastic," *Environmental History* 2, no. 3 (July 1997): 280–281. Mailer was speaking specifically about plastics.

95. Grover C. Mueller, "Coal from Cabbages," *Popular Science Monthly* (February 1929): 18 (accessed August 28, 2013); "Miners Eat Coal to Survive," *Chicago Tribune*, August 29, 2007 http://articles.chicagotribune.com/2007–08–29/news/0708290725_1_coal-trapped-deadliest (accessed August 28, 2013); "Pregnancy Cravings on the Rise," *BBC News*, April 28, 2008, http://news.bbc.co.uk/2/hi/health/7370524.stm (accessed August 28, 2013).

96. MacCleery, *American Forests*, 4.

97. See Chapter 1, n. 72. Using different assumptions, Catton argues that the quantity of fuel and work that fossil fuels provided by 1970 would have required (if converted back to biofuels) approximately twenty good acres of farmland per citizen to make up for the twenty fossil-fuel acres we depend on to elevate what he calls our phantom carrying capacity, 45.

98. Acreage put into production increased steadily up through the 1940s, after which the logic of intensification allowed for a decline (temporarily) in harvested acres. For harvested acreage, see US Department of Commerce, Bureau of the Census, "Series K 496–501: Acreages of Harvested Crops, by Use, and Indexes of Cropland Used for Crops and Crop Production per Acre: 1910 to 1970," *Historical Statistics of the United States*, 510.

99. US Department of Agriculture, National Agricultural Statistics Service, "Poultry Production and Value 2011 Summary" (April 2012), 5, http://usda .mannlib.cornell.edu/MannUsda/viewDocumentInfo.do?documentID=1130 (accessed February 25, 2013).

100. US Department of Agriculture, National Agricultural Statistics Service, "Livestock Slaughter 2011 Summary" (April 2012), 5, http://usda.mannlib .cornell.edu/MannUsda/viewDocumentInfo.do?documentID=1097 (accessed February 25, 2013).

101. Michael Pollan, "Power Steer," *New York Times Magazine*, March 31, 2002, 46–47; and Eric Schlosser, *Fast Food Nation: The Dark Side of the American Meal* (New York: Houghton Mifflin, 2001), 139–144.

102. Nikiforuk, *The Energy of Slaves*, 85. By 2007, that figure had climbed to nearly 16 percent, according to Patrick Canning, Ainsley Charles, Sonya Huang et al., "Energy Use in the US Food System," in *Energy Use in the U.S. Food System/ ERR-9,4* US Department of Agriculture (March 2010), iv.

103. Anthony Glass gives the figure of 100 billion kilograms of annual artificial nitrogen use worldwide, upwards of 100 million tons. Glass, "Nitrogen Use Efficiency of Crop Plants: Physiological Constraints upon Nitrogen Absorption," *Critical Reviews in Plant Sciences* 22, no. 5 (September 2003): 453. See also Vaclav Smil, "Nitrogen and Human Food Production: Proteins for Human Diets," *Ambio* 31, no. 2 (March 2002): 127.

104. For the story of nitrogen fertilizers, see Hugh Gorman's *The Story of N: A Social History of the Nitrogen Cycle and the Challenge of Sustainability* (New Brunswick, NJ: Rutgers University Press, 2013).

105. Michael Pollan, *Omnivore's Dilemma: A Natural History of Four Meals* (New York: Penguin, 2006), 32–37.

106. "The EPA and Food Security," http://www.epa.gov/pesticides/factsheets /securty.htm (accessed February 27, 2013); Martin C. Heller and Gregory A. Keoleian, "Assessing the Sustainability of the US Food System: A Life Cycle Perspective," *Agricultural Systems* 76 (2003): 1019, 1022, 1032.

107. The figures provided are from geologist Earl Ferguson Cook, "The Flow of Energy in an Industrial Society," *Scientific American* 225, no. 3 (1971): 136.

108. David Pimentel and Marcia Pimentel, *Food, Energy, Society* (New York: CRC Press, 2008), 6.

109. Heller and Keoleian, "Assessing the Sustainability of the US Food System," 1033.

110. Nikiforuk, *Energy of Slaves*, 85, 211.

111. McNeill, *Something New under the Sun*, xxiii, 4.

2. Rocks and Bodies

1. Chester C. Gilbert and Joseph Ezekiel Pogue, *Power: Its Significance and Needs*, Smithsonian Institution, US National Museum Bulletin 102, pt. 5 (Washington, DC: Government Printing Office, 1918), 7, 14.

2. Gilbert and Pogue's calculations are difficult to replicate because their assumption of 150 million horsepower units of consumption in the United States is not cited. Their admittedly "rough" figure can be better derived from current estimates. A more convincing argument, based on current estimates of horsepower output in 1920, is that Americans had available to them about 430 million mechanical horsepower units, or the equivalent of about 4 horsepower units per person (for a population of 106 million). Whereas Gilbert and Pogue estimate one horsepower unit to be twenty times the work capacity of an adult male, most scholars estimate one horsepower unit to be about six to ten times that capacity, a revised total output and ratio that would yield twenty-four to forty energy slaves per capita circa 1920.

The very big caveat to this statement is that only some of those mechanical slaves were being called into action at any given point in the day. At night most cars and steam engines were dormant, and there never was a time when Americans called upon all of their prime movers simultaneously. However, these figures give us an estimate of the army of potential energy slaves available for use at any given time. In an organic world, they would have had to have been fed, clothed, warmed, and sheltered whether they were in use or not.

To complicate matters, Gilbert and Pogue were writing at a critical turning point in the nation's use of power, before automobiles, trucks, and buses were fully a part of the equation. If we push the story forward to 1930, the nation's available energy in mechanical prime movers had skyrocketed to a bewildering 1.6 billion horsepower units, or the equivalent in work potential, using their logic, of about 32 billion manpower units, again only some of which was being activated at any given moment. Whatever the case, their basic point holds true: Americans would have needed a very large quantity of literal manpower (and a lot of consequent slack in the ecosystem) in order to meet their daily power requirements under the old terms by the 1920s. See ibid.; US Department of Commerce, Bureau of the Census, "Series S 1–14: Total Horsepower of All Prime Movers: 1849–1970," *Historical Statistics of the United States*, 818.

3. Gilbert and Pogue, *Power*, 14.

4. Theodore R. Schatzki and Wolfgang Natter, "Sociocultural Bodies, Bodies Sociopolitical," in *The Social and Political Body*, ed. Theodore R. Schatzki and Wolfgang Natter (New York: Guilford Press, 1996), 5–10.

5. See, for example, Peter Hitchcock, "Oil in an American Imaginary," *New Formations* 69 (Summer 2010): 81–97.

6. Lemenager, "The Aesthetics of Petroleum, after *Oil!*" *American Literary History* 24, no. 1 (Spring 2012): 60.

7. Schatzki and Natter, "Sociocultural Bodies, Bodies Sociopolitical," 5.

8. "Energy, n.," *Oxford English Dictionary Online* (Oxford, UK: Oxford University Press, June 2013), http://www.oed.com.ezproxy.nu.edu/view/Entry

/62088?isAdvanced=false&result=1&rskey=GoxYiF& (accessed August 03, 2013).

9. This calculation is a ballpark figure that assumes a 280-pound yield of meat from a 1,000-pound steer. The energy density of each pound of meat is calculated at approximately 1,000 calories/pound, or a total of 280,000 calories. The energy density in coal is calculated at approximately 3,150 calories/pound. For comparative energy densities and efficiencies, see Smil, *Energy in World History* (Cambridge: Massachusetts Institute of Technology Press, 1999), xvi.

10. See US Department of Commerce, Bureau of the Census, "Series S-14: Total Horsepower of All Prime Movers: 1849–1970," *Historical Statistics of the United States*, 818.

11. The term *the human motor* is taken from Anson Rabinbach, *The Human Motor: Energy, Fatigue, and the Origins of Modernity* (Berkeley: University of California Press, 1992).

12. Physiologists and engineers could now argue, for instance, that the human motor was only about 15 percent energy efficient in turning food calories into usable labor, whereas a diesel engine might turn petroleum calories into work at the much higher rate of 30 percent or so. Smil, "Table 7: Efficiencies of Common Energy Conversions," in *Energies: An Illustrated Guide to the Biosphere and Civilization* (Cambridge: Massachusetts Institute of Technology Press, 1999), xvi.

13. Rabinbach, *The Human Motor*, 72–81.

14. As an example of Taylor's efforts to discipline the human body to a machine ethic, we might turn to the example of the pig-iron handler in Frederick Winslow Taylor, *The Principles of Scientific Management* (Charleston, SC: Biblio-Bazaar Reproduction Series, 1997), 35–38.

15. *The River*, directed by Pare Lorentz (1938; Farm Security Administration), http://archive.org/details/TheRiverByPareLorentz.

16. Andrew Nikiforuk, *The Energy of Slaves: Oil and the New Servitude* (Vancouver, BC: Greystone Books, 2012), 90.

17. Stuart Chase, *Men and Machines* (New York: Macmillan, 1929), 37.

18. See US Department of Commerce, Bureau of the Census, "Series A29–42: Annual Estimates of the Population, by Age: 1900–1970," *Historical Statistics of the United States*, 10; "Series 1–14: Total Horsepower of All Prime Movers: 1849–1970," *Historical Statistics of the United States*, 818.

19. Richard White, *The Organic Machine: The Remaking of the Columbia River* (New York: Hill and Wang, 1992), 3, 7.

20. William Catton, "Man, the Prosthetic Animal," in *Overshoot: The Ecological Basis of Revolutionary Change* (Urbana: University of Illinois Press, 1980), 143–156; and Lemenager, "Aesthetics of Petroleum," 70.

21. Lemenager, "Aesthetics of Petroleum," 62.

22. Lewis Mumford, *Technics and Civilization* (New York: Harcourt and Brace, 1963), 144.

23. Charles Lindbergh, *The Spirit of Saint Louis* (1953; New York: Scribner, 2003), 353.

24. See Chapters 1, 5, 6, and 8, Stephen Kern, *The Culture of Time and Space, 1880–1920* (Cambridge, MA: Harvard University Press, 1983).

25. Chase, *Men and Machines*, 144.

26. Mumford, *Technics and Civilization*, 199, 236.

27. Ibid. See also Stuart Chase, *Prosperity: Fact or Myth* (New York: Charles Boni, 1929), 45.

28. Kenneth Fearing, "Dirge," in Fearing, *Collected Poems of Kenneth Fearing* (New York: AmPress, 1940), 60; William Carlos Williams, "Spring and All," in *The Collected Poems of William Carlos Williams: 1909–1939*, ed. A. Walton Litz and Christopher MacGowen (New York: New Directions, 1991), 205–206.

29. Lemenager, "Aesthetics of Petroleum," 27.

30. Katherine A. Fisher, "It's Brain Not Brawn That Runs a Modern Kitchen," *Good Housekeeping* 91, no. 1 (July 1930): 78–79.

31. Chase, *Men and Machines*, 9

32. David Nye, *Consuming Power: A Social History of American Energies* (Cambridge: Massachusetts Institute of Technology Press, 1999), 250.

33. *Mother Takes a Holiday*, produced for Whirlpool Company (1952; Jam Handy Productions), http://archive.org/details/MotherTa1952.

34. For a sharp discussion of the tempo of modern life, see Chapter 1, Joel Dinerstein, *Swinging the Machine: Modernity, Technology, and African American Culture between the World Wars* (Amherst, MA: University of Massachusetts Press, 2003), 29–62.

35. Henry Ford, *My Life and Work* (Garden City, NY: Doubleday, Page, 1922), 26, 159. See Ford's description of men and machines in Chapter 103, "Terror of the Machine," 101–115.

36. The argument that the assembly line aimed to reduce workers to a raw "physical animality" is Antonio Gramsci's. The validity and limits to that argument are discussed in Ernest J. Yanarelli and Herbert G. Reid, "From 'Trained' Gorillas to 'Humanware': Repoliticizing the Body-Machine Complex between Fordism and Post-Fordism," in *The Social and Political Body*, ed. Theodore R. Schatzki and Wolfgang Natter (New York: Guilford Press, 1996), 181–229. To be sure, fossil fuels, in and of themselves, were not responsible for cramping the subaltern body or for taking the creativity and joy out of work. Modernity's energies were channeled through already well-worn paths. Much of the world's industrial and agricultural labor had always been repetitive and unpleasant, and much of the play and excess of preindustrial labor had already been eliminated from the nation's own industries, farms, and slave plantations for reasons that had nothing to do with coal, oil, or natural gas. As Chase explained, the feeling of being a slave to the machine found its parallel in somatic civilizations that had their own share of galley slaves and hewers of wood whose lives were comprised, as he put it, of "dull, hard, debasing work." In *Men and Machines*, 168.

37. Ford, *My Life and Work*, 93, 108.

38. See Chapter 6, Charles Rumford Walker, *The Man on the Assembly Line* (Cambridge, MA: Harvard University Press, 1952), 87–88.

39. David Nye, *America's Assembly Line* (Cambridge: Massachusetts Institute of Technology Press, 2013), 102.

40. See oral interview with Paul Boatin in *The Great Depression: A Job at Ford's*, directed by Jon Else (WGBH Boston, 1993).

41. Ford, *My Life and Work,* 103.

42. *Modern Times,* directed by Charles Chaplin (1936; Image Entertainment 2000), DVD.

43. *Muller v. Oregon*—208 U.S. 412 (1908). WestlawNext.

44. Ibid.

45. See also John L. Lewis, "Labor and the Nation," in *The World's Greatest Speeches,* ed. Lewis Copeland et al. (Mineola, NY: Dover, 1999), 416.

46. Nye, *Consuming Power,* 71–83, 140–147.

47. There were 693,800 coal miners in 1919, plus other mining employees, out of an adult male labor force of approximately 32 million; US Department of Commerce, Bureau of the Census, *Historical Statistics of the United States,* 131, 580.

48. Earl Brown, interview with Cliff Kuhn, June 29, 1984, Working Lives Oral History Project, University of Alabama, http://acumen.lib.ua.edu/u0008 _0000003_0000008 (accessed August 21, 2013).

49. Ibid.

50. Ibid.

51. Thomas Andrews discusses the concept of the "workscape." See the intimate sensory regime of the coal mine in Andrews, *Killing for Coal: America's Deadliest Labor War* (Cambridge, MA: Harvard University Press, 2008), 144–145, 167.

52. Larry Mantowich, interviewed by Bobbie Herndon, 1974, Oral History Office, Sangamon State University, 24. Available in Norris L. Brookens Library, University of Illinois at Springfield, M319, http://library.uis.edu/archives /collections/oral/pdf/MANTOWICHvI.pdf (accessed August 21, 2013).

53. Jack Battuello interviewed by Nick Cherniavsky and Bobbie Herndon, 1973, Oral History Office, Sangamon State University. Available in Norris L. Brookens Library, University of Illinois at Springfield, M319, http://www.ida illinois.org/cdm/ref/collection/uis/id/967 (accessed February 25, 2013).

54. Reverend Joe Freeman, "There Will Be No Black Lung (Up in Heaven)," *Music of Coal: Mining Songs from the Appalachian Coalfields* (Lonesome Records and BMI Publishing, 2007), 47.

55. Merle Travis, "Dark as a Dungeon," in ibid.,15.

56. Mantowich interview, 27.

57. Nye, *Consuming Power,* 88.

58. Mantowich interview, 26.

59. "Dream of a Miner's Child," in *Music of Coal,* 23.

60. Ibid.

61. *How Green Was My Valley,* directed by John Ford (1941; 20th Century Fox, 2003), DVD.

62. Lewis Mumford, *The City in History: Its Origins, Its Transformations, and Its Prospects* (New York: Harcourt Brace Jovanovich, 1961), 450.

63. Sherwood Anderson, *Puzzled America* (Mamaroneck, NY: Paul P. Appel, 1970), 58.

64. Luis Fernández-Galiano, *Fire and Memory: On Architecture and Energy* (Cambridge: Massachusetts Institute of Technology Press, 2000), 63–67.

65. Amrita Dosanjh, "Childhood Asthma and Anthropogenic CO_2 Emissions," *Journal of Asthma and Allergy* 4 (2011): 105; David L. Buckeridge, Richard

Glazier, Bart J. Harvey et al., "Effect of Motor Vehicle Emissions on Respiratory Health in an Urban Area," *Environmental Health Perspectives* 110, no. 3 (March 2002): 293.

66. Garret Keizer, *The Unwanted Noise of Everything We Want* (New York: Public-Affairs, 2010), 31–32.

67. Mumford, *Technics and Civilization*, 178–179.

68. Dinerstein, *Swinging the Machine*, 12, 71, 261.

69. Cecelia Tichi, *Shifting Gears: Technology, Literature, Culture in Modernist America* (Chapel Hill: University of North Carolina Press, 1987), xv, 41, 72.

70. Tichi, *Gears and Girders*, 18, 31, 33, 52–53.

71. Paul Sutter, "The World with Us: The State of Environmental History," *Journal of American History* 100, no. 3 (2013): 94–119.

3. An Upthrust into Barbarism

1. Robert Bruère, "The Coming of Coal," *Survey Graphic*, March 25, 1922, 979–983.

2. Thomas Andrews, *Killing for Coal: America's Deadliest Labor War* (Cambridge, MA: Harvard University Press, 2008), 54.

3. Bruère, "The Coming of Coal," 982–983.

4. Dean Jensen, "Burchfield's *Plumbed the Soul*," *Milwaukee Sentinel*, April 27, 1984, 20.

5. Coal was until the middle of the twentieth century the dominant fossil fuel in the United States, supplying approximately 25 percent of US energy needs in 1870 and peaking at 72 percent in 1920. Thereafter, US reliance on petroleum and natural gas surpassed coal in terms of percentages, although the nation continued to increase its overall use of coal. This is accounted for by an overall increase in energy consumption. At the end of the twentieth century, coal was still supplying nearly 50 percent of the energy required to generate electricity in the United States, and 80 percent of US energy needs were dependent on fossil fuels of one type or another. See US Department of Energy, "Energy in the United States: A Brief History and Current Trends," http://www.eia.doe.gov/emeu/aer/eh1999/eh1999.html (accessed July 26, 2001). See also Louis Hunter's *A History of Industrial Power in the United States*, vol. 2: *Steam Power* (Cambridge: Massachusetts Institute of Technology Press, 1991), 430–431.

6. Stephen Mosley has argued that in the European context coal in the industrial period called up two opposing narratives—the one pointing to coal and its smoke as the "barbarous signs" of modern life; the other equating coal and smoke with the "familiar and reassuring" signs of progress. Mosley, *The Chimney of the World: A History of Smoke Pollution in Victorian and Edwardian Manchester* (Cambridge, UK: White Horse Press, 2001).

7. Rebecca Harding Davis, *Life in the Iron Mills; or, the Korl Woman* (New York: Feminist Press, 1972), 1.

8. Sherwood Anderson, "My Fire Burns," *Survey Graphic*, March 25, 1922, 999.

9. Lewis Mumford, *Technics and Civilization* (New York: Harcourt and Brace, 1963), 371; Gifford Pinchot, "Introduction," *Survey Graphic*, March 1, 1924, 1; Charles Steinmetz, "The White Revolution," *Survey Graphic*, October 1, 1923, 2; Joseph K. Hart, "Power and Culture," *Survey Graphic*, March 1, 1924, 625–628.

10. Anderson, "My Fire Burns," 998.

11. Ibid., 999–1000.

12. Ibid.

13. Ibid., 999; Hart, "Power and Culture," 626.

14. Peter Balbert, "Gripped by the Ultimate Master: Mining, Manhood, and Morbidity in D. H. Lawrence's 'Odour of Chrysanthemums,'" in *Caverns of the Night: Coal Mines in Art, Literature, and Film*, ed. William Thesing (Columbia: University of South Carolina Press, 2000), 89.

15. Winthrop Lane, "The Black Avalanche," *Survey Graphic*, March 25, 1922, 1006.

16. Ibid., 1002.

17. Ibid., 1003–1004; Winthrop Lane, *Civil War in West Virginia: A Story of the Industrial Conflict in the Coal Mines* (New York: B. W. Huebsch, 1921), 21–30.

18. Arthur Gleason, "Company-Owned Americans," *Nation*, June 12, 1920, 794.

19. Lane, "Black Avalanche," 1004.

20. Lewis Hine, "The New West Virginians," *Survey Graphic*, March 25, 1922, 1005.

21. George Leighton, "Shenandoah, Pennsylvania: The Story of an Anthracite Coal Town," *Harper's*, January 1937, 147.

22. Anderson, "My Fire Burns," 997.

23. See Chapter 6 in David Nye, *America as Second Creation: Technology and Narratives of New Beginnings* (Cambridge: Massachusetts Institute of Technology Press, 2003).

24. Crane, "In the Depths of a Coal Mine," *McClure's* (1894): 195–209.

25. See discussion of the irrational, the repressed, and the "low" in Peter Stallybrass and Allon White, *The Politics and Poetics of Transgression* (Ithaca, NY: Cornell University Press, 1986), 2–5.

26. Crane, "In the Depths of a Coal Mine," 209.

27. Ibid., 200.

28. Patrick Dooley, "Openness to Experience in Stephen Crane's 'In the Depths of a Coal Mine,'" *Caverns of the Night: Coal Mines in Art, Literature, and Film*, ed. William Thesing (Columbia: University of South Carolina Press, 2000), 192.

29. Joseph Katz, "Stephen Crane: Muckraker," *Columbia Library Column* 17, no. 2 (July 1968): 5.

30. Crane, "In the Depths of a Coal Mine," 209.

31. There is a lively debate as to whether *McClure's* editors exercised censorship in publishing this piece (as Crane's companion Linson explained) or whether Crane himself edited the piece for aesthetic reasons (as some later biographers have suspected). See Katz and Dooley.

32. Crane, "In the Depths of a Coal Mine," 207–209.

33. William Gilmore Beymer, "In the Earth Beneath," *Harper's Monthly* (October 1909): 775.

34. Ibid., 774.

35. Ibid., 775.

36. Ibid., 775, 780.

37. Ibid., 775, 781, 782–785.

38. Ibid., 775.

39. Eugene O'Neill, *The Hairy Ape*, in *Eugene O'Neill Complete Plays, 1920–1931* (New York: Modern Library, 1937), 200, 222.

40. Ibid., 198.

41. Ibid., 210, 198.

42. Ibid., 186.

43. Ibid., 185.

44. Joyce Flynn, "Sites and Sights: The Iconography of the Subterranean in Late Nineteenth-century Irish-American Drama," *Melus* 18, no. 1 (1993): 6–7.

45. O'Neill, *Hairy Ape*, 201.

46. O'Neill, *Hairy Ape*, 203; Patrick Chura, "'Vital Contact': Eugene O'Neill and the Working Class," *Twentieth-Century Literature* 49, no. 4 (2003): 533.

47. O'Neill, *Hairy Ape*, 209, 213.

48. Ibid., 214, 210; Chura, "'Vital Contact,'" 533.

49. See, for instance, Brian Kelly, "Having It Their Way: Alabama Coal Operators and Their Search for Docile Labor, 1908–1921" in *It Is Union and Liberty: Alabama Coal Miners and the UMW*, ed. Edwin Brown (Tuscaloosa: University of Alabama Press, 1999), 38–61.

50. Although Herbert Gutman has seen in the United Mine Workers (UMW) a model of interracial solidarity and participatory democracy, others have pointed out that the UMW may have been less democratic than Gutman supposed. See "The Negro and the United Mine Workers of America: The Career and Letters of Richard L. Davis and Something of Their Meaning, 1890–1900," in Herbert Gutman, *Work, Culture, and Society in Industrializing America* (New York: Knopf, 1976).

51. Carl Sandburg, "Sayings of Henry Stephens," *Survey Graphic*, March 25, 1922, 1044.

52. Balbert, "Gripped by the Ultimate Master," 89.

53. Ibid.

54. *The Proud Valley* directed by Pen Tennyson (1940; Cozumel Films, 2004), DVD.

55. Bosley Crowther, "The Screen: A Story of Welsh-Singing Miners," *New York Times*, May 17, 1941, 19.

56. Thomas Cripps, "Paul Robeson and Black Identity in American Movies," *Massachusetts Review* 11, no. 3 (Summer 1970): 468–485.

57. O'Neill, *Hairy Ape*, 198.

4. The Dynamo-Mother

1. Eugene O'Neill, *The Dynamo*, in *Eugene O'Neill: Complete Plays, 1920–1931* (New York: Library of America, 1988), 884.

2. Eugene O'Neill, letter to Michael Gold, August 15, 1928, in *Selected Letters of Eugene O'Neill*, ed. Travis Bogard and Jackson R. Bryer (New Haven, CT: Yale University Press, 1998), 308.

3. David Nye, *American Technological Sublime* (Cambridge: Massachusetts Institute of Technology Press, 1994), 143–172.

4. Carolyn de la Pena, *The Body Electric: How Strange Machines Built the Modern American* (New York: New York University Press, 2005), 137–170.

5. Charles Steinmetz, "The White Revolution," *Survey Graphic*, October 1, 1923, 1035; and Thomas Hughes, *American Genesis: A Century of Innovation and Technological Enthusiasm, 1870–1970* (Chicago: University of Chicago Press, 2004), 2, 473.

6. Lewis Mumford, *Technics and Civilization* (New York: Harcourt and Brace, 1963), 151–153, 212.

7. Gifford Pinchot, "Giant Power," *Survey Graphic*, March 1, 1924, 561–562. See also Pinchot, "Introduction," *Annals of the American Academy of Political and Social Science* 118 (March 1925): vii–xii.

8. Hart, "Power and Culture," 625–628; and Mumford, *Technics and Civilization*, 151–211.

9. Hart, "Power and Culture," 625–628.

10. Stuart Chase, *The Promise of Power* (New York: John Day, 1933), n.p.

11. Ibid.; Pinchot, "Giant Power," 561–562.

12. Chase, *Promise of Power*, n.p.

13. Ibid.

14. Paul Kellogg, "The Play of a Big Man with a Little River," *Survey Graphic*, March 1, 1924, 637–642, 658–664; James C. Scott, *Seeing Like a State: How Certain Schemes to Improve the Human Condition Have Failed* (New Haven, CT: Yale University Press, 1998).

15. Chase, *Promise of Power*, n.p.

16. Ibid.

17. Hart, "Power and Culture," 625–628.

18. W. S. Murray, "The Superpower System as an Answer to a National Power Policy," *General Electric Review* 25 (February 1922): 72–4.

19. Ibid., 173. The criticisms lodged against the superpower scheme were, in good part, fair. Superpower's main proponents had fought quite vocally to make sure that the hand of government would remain light after the system was designed and authorized. That premise of corporate control over the nation's power (with the very loose local oversight of the utilities) was a matter of deep concern for progressive critics who saw public control of electricity as the most pressing issue of the 1920s. In fact, the principle of government development and regulation of the nation's energy and physical power was the key plank of the Progressive Party for bringing literal power (and with it social power) back to the people.

20. John Dos Passos, *The Big Money* (New York: Harcourt, Brace, 1936), 528–529.

21. Pinchot, "Introduction," *Annals*, xi.

22. For contemporaneous interpretations of the Giant Power Survey, see the series of articles in both: "Giant Power: Special Edition," *Survey Graphic*, March 1,

1924; and "Giant Power: Large-Scale Electrical Development as a Social Factor," *Annals of the American Academy of Political and Social Science* 118 (March 1925). For secondary scholarship on the Giant Power Survey, see Jean Christie, "Giant Power: A Progressive Proposal of the 1920s," *Pennsylvania Magazine of History and Biography* 96 (October 1972): 480–507; Leonard DeGraff, "Corporate Liberalism and Electric Power System Planning in the 1920s," *Business History Review* 64 (Spring 1990): 1–31; and Sarah Phillips, *This Land, This Nation: Conservation, Rural America, and the New Deal* (New York: Cambridge University Press, 2007).

23. Pinchot, "Introduction," *Annals*, xi.

24. Mumford, *Technics and Civilization*, 215, 224, 254–255.

25. The phrase "civilizing the machine" comes from John Kasson, *Civilizing the Machine: Technology and Republican Values in America, 1776–1900* (New York: Hill and Wang, 1999).

26. George Norris, quoted in Carl D. Thompson, *Confessions of the Power Trust: A Summary of the Testimony Given in the Hearings of the Federal Trade Commission on Utilities Corporations Pursuant to Resolution No. 83 of the United States Senate Approved February 15, 1928* (New York: Dutton, 1932), xvii. Reprinted in 1933.

27. Pinchot, quoted in ibid., xvii–xviii.

28. Thompson, *Confessions of the Power Trust*, xviii.

29. Ibid., 19–20.

30. Ibid., 13–14, 20, 25–26.

31. Ibid., 270–273.

32. Ibid., 101, 309–329.

33. Ibid., 295–308.

34. Ibid., 269, 282.

35. David Nye, *Electrifying America: Social Meanings of a New Technology, 1880–1940* (Cambridge, MA: Massachusetts Institute of Technology Press, 1992), 340.

36. The New Deal electrification programs have been studied intensely for their political, social, and cultural importance. The most important works on the subject include Finis Dunaway, *Natural Visions: The Power of Images in American Environmental Reform* (Chicago: University of Chicago Press, 2008); Sarah Phillips, *This Land, This Nation*; David Nye, *Electrifying America*; and Ronald C. Tobey, *Technology as Freedom: The New Deal and the Electrical Modernization of the American Home* (Berkeley: University of California Press, 1996).

37. The New Deal Network has reproduced this Living Newspaper production in its entirety with commentary at http://newdeal.feri.org/power/index .htm (accessed March 31, 2012).

38. Ibid.

39. Ibid.

40. Ibid.

41. Ibid.

42. Ibid.

43. Ibid.

44. Ibid.

45. Tobey, *Technology as Freedom*.

46. Jürgen Martschukat, "'The Art of Killing by Electricity': The Sublime and the Electric Chair," *Journal of American History* 89, no. 3 (December 2003).

47. Tony Long, "Jan. 4, 1903: Edison Fries an Elephant to Prove His Point," *Wired*, January 4, 2008, http://www.wired.com/science/discoveries/news/2008/01/dayintech_0104 (accessed March 31, 2012).

48. William Hard, "Giant Negotiations for Giant Power: An Interview with Herbert Hoover," *Survey Graphic*, March 1, 1924, 577.

49. See, for instance, Rhitu Chaterjee, "Hazardous Waste: TVA Spill's Chemical Legacy," *Environmental Health Perspectives* 117, no. 8 (August 2009): A346; Winona LaDuke, "TVA's Big Plans for Navajo Uranium," *Business and Society Review* (March 1, 1979): 22–25; and the considerable news coverage in the 1960s and 1970s of the TVA's environmental practices with regard to strip mining, air pollution, and nuclear use.

50. O'Neill, *Dynamo*, 884.

51. For a contemporary description of the human and environmental costs of those dependencies, see Jeff Goodell, *Big Coal: The Dirty Secret behind America's Energy Future* (New York: Houghton Mifflin, 2006).

5. A Faint Whiff of Gasoline

1. *Giant*, directed by George Stevens (1956; Warner Home Video, 2005), DVD.

2. Stephanie Lemenager contends that oil has also been associated with the lower strata of the body, with, as she says, "shit" and "filth." We have already seen this association played out in relationship to coal. Lemenager, "The Aesthetics of Petroleum, after *Oil!*" *American Literary History* 24, no. 1 (Spring 2012): 73.

3. Petroculture, as a term, has increasingly entered into circulation as the result of the McGill Institute's annual conference by that name, codirected by Imre Szeman and Sheena Wilson.

4. See Worden's interesting claim that Steven's film was part of a larger cultural tendency in the national imaginary to differentiate between bad oil and good oil cultures mostly by "repressing" the industry and its practices in the latter. Daniel Worden, "Fossil-Fuel Futurity: Oil in Giant," *Journal of American Studies* 46 (2012): 443, 445.

5. Stevens, *Giant*.

6. Thomas Hughes, *American Genesis: A Century of Innovation and Technological Enthusiasm, 1870–1970* (Chicago: University of Chicago Press, 2004), 473n2.

7. Stuart Chase, "Gasless America," *Nation*, December 8, 1926, 586–587.

8. Peter Hitchcock, "Oil in an American Imaginary," *New Formations* 69 (Summer 2010): 81.

9. Chase, "Gasless America," 586–587; Lewis Mumford, *Technics and Civilization* (New York: Harcourt and Brace, 1963), 235.

10. Charles Palmer, film script for *Fire and the Wheel*, produced for General Petroleum Corporation, a Socony-Mobil Company (Parthenon Pictures, ca. 1955). Exxon Mobil. Box 2.207/E82.

11. On the oil industry's internal responses to its bad press, see Roger Olien and Diana Olien, *Oil and Ideology: The Cultural Creation of the American Petroleum Industry* (Chapel Hill: University of North Carolina Press, 2000).

12. Lemenager, "Aesthetics of Petroleum," 27–28.

13. Upton Sinclair, *Oil!* (1927; New York: Penguin, 2007), 70.

14. Olien and Olien, *Oil and Ideology*, 98–100; Myrna Santiago, *The Ecology of Oil: Environment, Labor, and the Mexican Revolution, 1900–1938* (Cambridge: Cambridge University Press, 2006); Paul Sabin, *Crude Politics: The California Oil Market, 1900–1940* (Berkeley: University of California Press, 2005), 28–29, 46–49, 11–135; Daniel Yergin, *The Prize: The Epic Quest for Oil, Money, and Power* (New York: Free Press, 1992), 250–251. For contemporaneous discussion of the importance of energy to American life and the risk of its unfair expropriation by private interests, see Gifford Pinchot, "Address Delivered by Gifford Pinchot, Governor of Pennsylvania, before the Survey Associates, Waldorf Astoria Hotel, New York," February 25, 1924, *Survey Associates Records*, Box 101, Folder 769. For a description of the poor conservation practices, see Charles Beard, *The American Leviathan: The Republic in the Machine Age* (New York: Macmillan, 1930), 546–549; Stuart Chase, *The Tragedy of Waste* (New York: Macmillan, 1937), 237, 251.

15. See, for instance, Chapter 4, "Believing the Worst," in Olien and Olien, *Oil and Ideology*, 83–118.

16. Quoted in ibid., 24.

17. Ida Tarbell, "John D. Rockefeller: A Character Study, II," *McClure's* (August 1905): 386.

18. Robert Lynd, "Done in Oil," *Survey Graphic* (November 1922); Lynd, "Crude Oil Religion," *Harper's*, September 1922.

19. Lynd, "Done in Oil," 137.

20. Lynd, "Crude-Oil Religion," 425.

21. Lynd, "Done in Oil," 137, 140–141, 144–146.

22. Ibid.

23. Ibid., 140, 145.

24. Ibid., 138, 140–144.

25. Ibid.

26. Ibid., 140, 142, 145; Lynd, "Crude-Oil Religion," 426.

27. Lynd, "Done in Oil," 138, 142.

28. John D. Rockefeller, Jr. "A Promise of Better Days," *Survey Graphic*, November 1922, 147–148. See, for instance, Brian Black's descriptions of the socioeconomic and environmental disarray that followed the Pennsylvania oil boom in *Petrolia: The Landscape of America's First Oil Boom* (Baltimore, MD: Johns Hopkins University Press, 2000); Myrna Santiago's and Robert Vitalis's account of the US (and British) oil industry's exploitative racialized class structures that defined the Saudi and Vera Cruz fields. Santiago, *Ecology of Oil*, and Vitalis, *America's Kingdom: Mythmaking on the Saudi Oil Frontier* (Palo Alto, CA: Stanford University Press, 2007).

29. See also Charles Harvey, "Robert S. Lynd, John D. Rockefeller, and *Middletown*," *Indiana Magazine of History* 79, no. 4 (December 1983): 333–336; and

Staughton Lynd, "Making Middletown," *Indiana Magazine of History* 101, no. 3 (September 2005): 229–231. Lynd's son surmises that the Rockefeller Institute directorship his father received to study Muncie, Indiana, might have been a strategic effort on the Rockefeller's part for redirecting his father's criticisms of the industry: perhaps "the boss was," he writes, "dealing with an outspoken shop-floor militant by making the man a foreman."

30. Olien and Olien, *Oil and Ideology*, 171.

31. Thanks to Robert Lifset for directing me to the long list of films on oil and culture from oil's early history forward.

32. Lemenager, "Aesthetics of Petroleum," 63.

33. Lemenager sees in Sinclair's *Oil* a slightly more complicated relationship to oil that includes, as she puts it, an "emotional attachment" in the text to both driving (an oil-related activity) and even oil drilling itself, thus reflecting in the text these interesting strains in the modern relationship to oil. Ibid., 62–63, 70.

34. Ibid., 75.

35. Sinclair, *Oil!*, 158.

36. Ibid., 90, 99.

37. Ibid.

38. Ibid.

39. Ibid.

40. Ibid., 409.

41. *Louisiana Story*, directed by Robert Flaherty (1948; Home Vision Studio, 2003).

42. Although Flaherty's contract with Standard Oil of New Jersey granted him considerable artistic freedom, students of the film have often downplayed the contractual constraints under which Flaherty worked. The main contract between Flaherty and Standard Oil assures the company there will be no major deviations from the original script, which details the "benefits of oil to mankind." See Eva Orbanz, ed., *Filming Robert Flaherty's* Louisiana Story: *The Helen van Dongen Diary* (New York: Museum of Modern Art, 1998): 93–94; and also *Variety Review* quoted in the *American Film Institute Catalog* (Berkeley: University of California Press, 1971).

43. See Jack Coogan, "Louisiana Story and the Ecological Imagination," *Wide Angle* 20, no. 2 (1998): 58–69; Richard Barsam, *The Vision of Robert Flaherty: The Artist as Myth and Filmmaker* (Bloomington: Indiana University Press, 1988), 102, 109.

44. Ibid.

45. Ibid.

46. *The Story of Colonel Drake*, directed by Arthur Pierson, produced for American Petroleum Institute (1954; Film Counselors, Inc.). Exxon Mobil. Box 2.207/E87. Later reissued under the title *Born in Freedom*.

47. Ibid.

48. See, for instance, Chapters 2–6 in Black, *Petrolia*, and Chapters 2–4 in Santiago, *Ecology of Oil*.

49. American Petroleum Institute, *Born in Freedom*.

50. Ibid.

51. Advertisement, "Oil Movie of the Year: The Story of Colonel Drake," *Saturday Evening Post*, ca. 1954, n.p.

52. American Petroleum Institute, *Destination Earth*, produced by John Sutherland (1956), http://archive.org/details/Destinat1956.

53. Palmer, *Fire and the Wheel*, 3–6.

54. *To New Horizons*, model design by Norman Bell Geddes, produced for General Motors (Jam Handy Productions 1940), http://www.youtube.com/watch?v=eNlgfkEgnWA&feature=related (accessed on April 3, 2012).

55. Eric Foner develops the notion of a "consumer's freedom" in *The Story of American Freedom* (New York: Norton, 1999), 262–274.

56. American Petroleum Institute, *Destination Earth*.

57. Ibid.

58. *A Fable*, directed by Rolf Brandis, produced for Mobil Oil (1969). Exxon Mobil, Box 2.207/E87.

59. Ibid.

60. Ibid.

61. Ibid.

62. According to Mobil's internal correspondence, the film, which had 15 million international viewers, was intended to "encourage [a] wider view of trade and private investment on the part of developing countries, through the elimination of barriers of economic nationalism," the idea of which was a "request from affiliate managers for [a] film to combat growing economic nationalism." See Raymond D'Argenio, letter to BL Hale, January 14, 1969; "Background Material for Meeting on Mobil Public Affairs' 'Communication Mix,'" June 1971. Exxon Mobil, Box 2.207/E82.

63. Charles Pollak, internal memo to Mr. Schmertz, July 19, 1971. Exxon Mobil, Box 2.207/E82.

Conclusion. *A Return of the Repressed*

1. Seth Stevenson, "Coal Mines Hotter: Sex Sells, but Can GE Use Sex to Sell Coal?" *Slate*, May 31, 2005, http://www.slate.com/articles/business/ad_report_card/2005/05/coal_miners_hotter.html (accessed January 30, 2012).

2. Alberta Energy, "Alberta's Oil Sands," http://www.energy.alberta.ca/OilSands/791.asp (accessed August 1, 2013).

3. Andrew Nikiforuk, *Tar Sands: Dirty Oil and the Future of a Continent* (Vancouver, BC: David Suzuki Foundation, 2009), 14–15, 129–134, 148–149.

4. For a discussion of the current technologies, practices, and risks of shale-oil production, see Vickram Rao, *Shale Gas: The Promise and Peril* (Research Triangle Park, NC: Research Triangle Institute Press, 2012), xi.

5. Eric Reece, *Lost Mountain: A Year in the Vanishing Wilderness* (New York: Riverhead Books, 2006), 128.

6. See Intergovernmental Panel on Climate Change, "Figure 3.5: Projections and Model Consistency of Relative Changes in Runoff by the End of the

21st Century," *Fourth Assessment Climate Change 2007 Synthesis Report,* http://www.ipcc.ch/publications_and_data/ar4/syr/en/figure-3-5.html (accessed August 10, 2013).

7. The current estimate is the world will see an increase of .2 degrees Celsius per decade through the end of the century. See Intergovernmental Panel on Climate Change, "3.2 Projections of Future Changes in Climate," *Fourth Assessment Climate Change 2007 Synthesis Report,* http://www.ipcc.ch/publications_and_data/ar4/syr/en/figure-3-2.html (accessed August 10, 2013).

8. Shan Li, "Exxon Mobil Tops *Fortune* List of World's Most Profitable Companies," *Los Angeles Times,* July 9, 2013, http://www.latimes.com/business/money/la-fi-mo-exxon-mobil-fortune-profit-rankings-20130709,0,6323483.story (accessed August 1, 2013).

9. Agustino Fontevecchia, "The End of Exxon's Reign? PetroChina Now the Biggest Oil Producer," *Forbes,* March 29, 2013, accessed http://www.forbes.com/sites/afontevecchia/2012/03/29/the-end-of-exxons-rein-petrochina-now-the-biggest-oil-producer/ (August 15, 2013). See also Associated Press, "Exxon Mobil's 2012 Profits Just Shy of Breaking Global Record," *Huffington Post,* February 1, 2013, http://www.huffingtonpost.com/2013/02/01/exxon-mobil-profit-world-record_n_2598502.html (accessed August 15, 2013).

10. Stanley Reed, "A Retrenched BP Tops Profit Forecasts," *New York Times,* April 30, 2014, http://www.nytimes.com/2013/05/01/business/global/bps-4-2-billion-profit-beats-forecasts.html?_r=0 (accessed August 15, 2013); Fiona Harvey, "BP Makes Record Profit as Its Safety Record Comes under Spotlight Again," *Guardian,* April 30, 2013, http://www.nytimes.com/2013/05/01/business/global/bps-4-2-billion-profit-beats-forecasts.html?_r=0 (accessed August 15, 2013).

11. Peabody Energy, "2012 Annual Report," http://www.peabodyenergy.com/mm/files/Investors/Annual-Reports/PE-AR2012.pdf, 2–3 (accessed August 21, 2013).

12. Jonathan Stein, "Oil and Coal Have Spent $427 Million to Influence Campaign in 2008," *Mother Jones,* August 18, 2008, http://www.motherjones.com/mojo/2008/08/oil-and-coal-have-spent-427-million-influence-campaign-2008.

13. Eric Lipton and Clifford Krauss, "Fossil Fuel Industry Ads Dominate TV Campaign," *New York Times,* September 13, 2012, http://www.nytimes.com/2012/09/14/us/politics/fossil-fuel-industry-opens-wallet-to-defeat-obama.html?pagewanted=all&_r=0 (accessed August 15, 2013).

14. Woody Woodrun, "Friends of Coal Unveil Governor's Trophy," *Inside the Herd Magazine,* August 23, 2006, http://marshall.scout.com/2/560161.html (accessed January 13, 2012).

15. Dana Jennings, "Grab a Brew While They Face Death," *New York Times,* March 24, 2011, AR1.

16. There is a wealth of recent writing on this subject. For example, see Timothy Gardner, "American Climate Skeptic Soon Funded by Oil, Coal Firms," Reuters, June 28, 2011, http://www.reuters.com/article/2011/06/28/us-usa-climate-skeptic-idUSTRE75R2HD20110628 (accessed August 21, 2013).

17. Gavin Aronsen, "Big Coal: Inbreeding, Not Mining to Blame for Birth Defects," *Mother Jones,* July 12, 2011, http://motherjones.com/blue-marble

/2011/07/mountaintop-removal-inbreeding-coal-mining (accessed January 13, 2012).

18. Ann Mulkern, "Coal Ad Blitz Launches New Spot as Industry Sees Political Gains," *New York Times*, February 8, 2010, http://www.nytimes.com /gwire/2010/02/08/08greenwire-coal-ad-blitz-launches-new-spot-as-industry -se-49401.html (accessed November 15, 2012). For the human and environmental costs of coal, see Barbara Freese, *Coal: A Human History* (New York: Penguin, 2004); Jeff Goodell, *Big Coal: The Dirty Secret behind America's Energy Future* (New York: Houghton Mifflin, 2006); and Reece, *Lost Mountain*.

19. US Energy Information Agency, "What Are the Major Sources and Users of Energy in the United States?" (esp. "Table 1.3: Primary Energy Use by Source and Sector, 2012"), *Monthly Energy Review* (April 2013), http://www.eia.gov /energy_in_brief/article/major_energy_sources_and_users.cfm (accessed August 21, 2013).

20. US Energy Information Administration, "How Much Petroleum Does the U.S. Import and from Where?" (April 2013), http://www.eia.gov/tools/faqs /faq.cfm?id=727&t=6 (accessed August 21, 2013); US Energy Information Administration, "U.S. Imports by Country of Origin" (April 2013), http://www .eia.gov/dnav/pet/pet_move_impcus_a2_nus_epoo_imo_mbbl_m.htm (accessed August 21, 2013).

21. Andrew Nikiforuk, *The Energy of Slaves: Oil and the New Servitude* (Vancouver, BC: Greystone Books, 2012), 180–203; Michael Watts, "The Anatomy of a Petro-Insurgency in the Niger Delta," in *Crude Domination: An Anthropology of Oil*, ed. Andrea Behrends, Stephen P. Reyna, and Günther Schlee (New York: Berghahn, 2011), 49–70.

22. US Department of Commerce, Bureau of the Census, "Series M 107–122. Bituminous Coal-Employment Strikes, and Domestic Consumption by Consumer Class; and Coke Production: 1880–1970," *Historical Statistics of the United States*, 590–591.

23. Paulette Young, *Annual Coal Report 2011*, US Energy Information Administration, November 2012, vii, http://www.eia.gov/coal/annual/pdf/acr.pdf (accessed August 21, 2013).

24. See Chapter 3 in Goodell, *Big Coal*.

25. Rick Michal, "Fifty Years Ago in December: Atomic Reactor EBR-1 Produced First Electricity," *Nuclear News* (November 2001): 28–29.

26. Jacob Adelman and Chisaki Watanabe, "Japan Studies Ice Wall to Halt Radioactive Water Leaks," *Bloomberg News*, August 14, 2013, http://www.bloom berg.com/news/2013–08–13/japan-studies-ice-wall-to-halt-radioactive-water -leaks.html (accessed August 26, 2013).

27. International Energy Agency, *2012 Key World Energy Statistic* (2012), 6, http://www.iea.org/publications/freepublications/publication/kwes.pdf (accessed August 21, 2013); US Energy Information Administration, "Table 1.3: Primary Energy Use by Source, 2012," *Monthly Energy Review* (March 2012) http:// www.eia.gov/energy_in_brief/article/major_energy_sources_and_users .cfm (accessed August 21, 2013).

28. Leslie White, *The Science of Culture: A Study of Man and Civilization* (New York: Grove Press, 1949), 362.

29. Clifford Krauss, "As Ethanol Takes Its First Steps, Congress Proposes a Giant Leap," *New York Times*, December 18, 2007, http://www.nytimes.com/2007/12/18/washington/18ethanol.html?pagewanted=all (accessed August 26, 2013).

30. Deepak Rajagopal and David Zilberman, "Review of Environmental, Economic, and Policy Aspects of Biofuels," Policy Research Working Paper 4341, World Bank Development Research Group, September 2007, 17–18, 22. See also Steven Sexton, David Zilberman, Deepak Rajagopal, and Gal Hochman, "The Role of Biotechnology in a Sustainable Biofuel Future," *AgBioForum* 12, no. 1 (2009): 130–140, http://www.agbioforum.org (accessed August 1, 2013).

31. There is a wide literature and high-pitched debates about contemporary agricultural practices. See, for instance, *Fatal Harvest: The Tragedy of Industrial Agriculture*, ed. Andrew Kimbrell (San Rafael, CA: Foundation for Deep Ecology, 2002); Paul Conkin, *Revolution Down on the Farm: The Transformation of American Agriculture since 1929* (Lexington: University Press of Kentucky, 2008).

32. Lewis Mumford, *Technics and Civilization* (New York: Harcourt and Brace, 1963), 158.

Appendix. Energy and Power

1. Vaclav Smil, *Energy in World History* (Boulder, CO: Westview Press, 1994), 1–11.

2. E. A. Wrigley, *Continuity, Chance, and Change: The Character of the Industrial Revolution in England* (New York: Cambridge University Press, 1988), 34–67.

3. Andrew Nikiforuk, *The Energy of Slaves: Oil and the New Servitude* (Vancouver, BC: Greystone Books, 2012), 65.

4. Smil, *Energy in World History*, 6–11.

Selected Bibliography

Alberta Energy. "Alberta's Oil Sands: Facts and Statistics," http://www.energy
.alberta.ca/OilSands/791.asp. Accessed August 19, 2013.

Andrews, Thomas. *Killing for Coal: America's Deadliest Labor War.* Cambridge, MA:
Harvard University Press, 2008.

Balbert, Peter. "Gripped by the Ultimate Master: Mining, Manhood, and Morbidity in D. H. Lawrence's 'Odour of Chrysanthemums.'" In *Caverns of the
Night: Coal Mines in Art, Literature, and Film.* Edited by William Thesing, 89–
103. Columbia: University of South Carolina Press, 2000.

Barrett, Ross, and Daniel Worden. "Oil Culture: Guest Editors' Introduction."
Journal of American Studies 46 (2012): 269–272.

Barsam, Richard. *The Vision of Robert Flaherty: The Artist as Myth and Filmmaker.*
Bloomington: Indiana University Press, 1988.

Bederman, Gail. *Manliness and Civilization: A Cultural History of Gender and Race
in the United States, 1880–1917.* Chicago: University of Chicago Press, 1995.

Berry, Wendell. *The Unsettling of America.* San Francisco: Sierra Club, 1996.

Black, Brian. *Petrolia: The Landscape of America's First Oil Boom.* Baltimore, MD:
Johns Hopkins University Press, 2000.

Bronson, Rachel. *Thicker Than Oil: America's Uneasy Partnership with Saudi Arabia.*
Oxford, UK: Oxford University Press, 2006.

Buckeridge, David L., Richard Glazier, Bart J. Harvey, Michael Escobar, Carl
Amrhein, and John Frank. "Effect of Motor Vehicle Emissions on Respira-
tory Health in an Urban Area." *Environmental Health Perspectives* 110, no. 3
(March 2002): 293–300.

Buell, Frederick. "A Short History of Oil Cultures; or, the Marriage of Catastrophe and Exuberance." *Journal of American Studies* 46 (2012): 273–293.

Burke, Edmund. "The Big Story: Human History, Energy Regimes, and the Environment." In *The Environment and World History.* Edited by Edmund Burke III
and Kenneth Pomeranz, 33–53. Berkeley: University of California Press,
2009.

Canning, Patrick, Ainsley Charles, Sonya Huang, and Karen R Polenske. *Energy
Use in the U.S. Food System/ERR-94* (March 2010). http://www.ers.usda.gov
/media/136418/err94_1_.pdf.

Caruth, Cathy. *Unclaimed Experience: Trauma, Narrative, and History.* Baltimore,
MD: Johns Hopkins University Press, 1996.

Catton, William. *Overshoot: The Ecological Basis of Revolutionary Change.* Urbana:
University of Illinois Press, 1980.

Center for Sustainable Systems, University of Michigan. *U.S. Material Use Factsheet.* Publication CSS05–18 (2012). http://css.snre.umich.edu/css_doc
/CSS05-18.pdf.

Chase, Stuart. *Men and Machines*. New York: Macmillan, 1929.
———. *The Promise of Power*. New York: John Day, 1933.
———. *The Tragedy of Waste*. New York: Macmillan, 1937.
Chaterjee, Rhitu. "Hazardous Waste: TVA Spill's Chemical Legacy." *Environmental Health Perspectives* 117, no. 8 (August 2009): A346, http://www.ncbi.nlm.nih.gov/pmc/articles/PMC2721889/.
Christie, Jean. "Giant Power: A Progressive Proposal of the 1920s." *Pennsylvania Magazine of History and Biography* 96 (October 1972): 480–507.
Chura, Patrick. "'Vital Contact': Eugene O'Neill and the Working Class." *Twentieth-Century Literature* 49, no. 4 (2003): 520–546.
Clark, Gregory. *A Farewell to Alms: A Brief Economic History of the World*. Princeton, NJ: Princeton University Press, 2007.
Conkin, Paul. *Revolution Down on the Farm: The Transformation of American Agriculture since 1929*. Lexington: University Press of Kentucky, 2008.
Coogan, Jack. "Louisiana Story and the Ecological Imagination." *Wide Angle* 20, no. 2 (1998): 58–69.
Cook, Earl Ferguson. "The Flow of Energy in an Industrial Society." *Scientific American* 225, no. 3 (1971): 135–146.
Cripps, Thomas. "Paul Robeson and Black Identity in American Movies." *Massachusetts Review* 11, no. 3 (Summer 1970): 468–485.
Cronon, William. *Changes in the Land*. New York: Hill and Wang, 1983.
———. *Nature's Metropolis: Chicago and the Great West*. New York: Norton, 1992.
Davis, Lance. *American Economic Growth: An Economist's History of the United States*. New York: Harper and Row, 1972.
de la Pena, Carolyn. *The Body Electric: How Strange Machines Built the Modern American*. New York: New York University Press, 2005.
Debeir, Jean-Claude, Jean-Paul Deléage, and Daniel Hémery. *In the Servitude of Power: Energy and Civilization through the Ages*. Translated by John Barzman. London: Zed Books, 1986.
Deepak, Rajagopal, and David Zilberman, "Review of Environmental, Economic, and Policy Aspects of Biofuels" (September 1, 2007). World Bank Policy Research Working Paper 4341. http://ssrn.com/abstract=1012473.
DeGraff, Leonard. "Corporate Liberalism and Electric Power System Planning in the 1920s." *Business History Review* 64 (Spring 1990): 1–31.
Dinerstein, Joel. *Swinging the Machine: Modernity, Technology, and African American Culture between the World Wars*. Amherst: University of Massachusetts Press, 2003.
Donahue, Brian. *The Great Meadow: Farmers and the Land in Colonial Concord*. New Haven, CT: Yale University Press, 2007.
Dooley, Patrick. "Openness to Experience in Stephen Crane's 'In the Depths of a Coal Mine.'" In *Caverns of the Night: Coal Mines in Art, Literature, and Film*. Edited by William Thesing, 186–198. Columbia: University of South Carolina Press, 2000.
Dosanjh, Amrita. "Childhood Asthma and Anthropogenic CO_2 Emissions." *Journal of Asthma and Allergy* 4 (2011): 103–105, doi: 10.2147/JAA.S24565.

Dunaway, Finis. *Natural Visions: The Power of Images in American Environmental Reform.* Chicago: University of Chicago Press, 2008.

Fernández-Galiano, Luis. *Fire and Memory: On Architecture and Energy.* Cambridge: Massachusetts Institute of Technology Press, 2000.

Flynn, Joyce. "Sites and Sights: The Iconography of the Subterranean in Late Nineteenth-Century Irish-American Drama." *Melus* 18, no. 1 (1993): 5–19.

Freese, Barbara. *Coal: A Human History.* New York: Penguin, 2004.

Foucault, Michel. "Nietzsche, Genealogy, History." In *Language, Countermemory, Practice: Selected Essays and Interviews.* Edited and translated by Donald F. Bouchard, 139–164. Ithaca, NY: Cornell University Press, 1977.

Giberti, Bruno. *Designing the Centennial: A History of the 1876 International Exhibition in Philadelphia.* Lexington: University Press of Kentucky, 2002.

Gilbert, Chester C., and Joseph Ezekiel Pogue. *Power: Its Significance and Needs.* Smithsonian Institution, US National Museum, Bulletin 102, Part 5. Washington, DC: Government Printing Office, 1918.

Gilroy, Paul. *Black Atlantic: Modernity and Double Consciousness.* Cambridge, MA: Harvard University Press, 1993.

Glass, Anthony. "Nitrogen Use Efficiency of Crop Plants: Physiological Constraints upon Nitrogen Absorption." *Critical Reviews in Plant Sciences* 22, no. 5 (September 2003): 453, doi: 10.1080/07352680390243512.

Goodell, Jeff. *Big Coal: The Dirty Secret behind America's Energy Future.* New York: Houghton Mifflin, 2006.

Goodheart, Andrew. "The Myth of the Machine." *Design Quarterly* 155 (Spring 1992): 24–28.

Gorman, Hugh. *The Story of N: A Social History of the Nitrogen Cycle and the Challenge of Sustainability.* New Brunswick, NJ: Rutgers University Press, 2013.

Greene, Ann. *Horses at Work: Harnessing Power in Industrial America.* Cambridge, MA: Harvard University Press, 2008.

Gutman, Herbert. *Work, Culture, and Society in Industrializing America.* New York: Knopf, 1976.

Hartman, Geoffrey H. "On Traumatic Knowledge and Literary Studies." *New Literary History* 26, no. 3 (1995): 537–563.

Heilbroner, Robert L., and Aaron Singer. *The Economic Transformation of America: 1600–Present.* New York: Harcourt, Brace, Jovanovich, 1984.

Heller, Martin C., and Gregory A. Keoleian. "Assessing the Sustainability of the U.S. Food System: A Life Cycle Perspective." *Agricultural Systems* 76 (2003): 1007–1041, http://dx.doi.org/10.1016/S0308-521X(02)00027-6.

Hitchcock, Peter. "Oil in an American Imaginary." *New Formations* 69 (Summer 2010): 81–97.

Holley, I. B. "Blacktop: How Asphalt Paving Came to the Urban Environment." *Technology and Culture* 44, no. 1 (October 2003): 703–733.

Hughes, Thomas. *American Genesis: A Century of Innovation and Technological Enthusiasm, 1870–1970.* Chicago: University of Chicago Press, 2004.

Hunter, Louis. *A History of Industrial Power in the United States.* Volume 2: *Steam Power.* Cambridge: Massachusetts Institute of Technology Press, 1991.

Intergovernmental Panel on Climate Change. *Climate Change 2007: Synthesis Report — Contribution of Working Groups I, II, and III to the Fourth Assessment Report of the Intergovernmental Panel on Climate Change.* Edited by Core Writing Team, R. K. Pachauri, and A. Reisinger. Geneva: IPCC, 2007.

International Energy Agency. *2012 Key World Energy Statistics.* Paris 2012. http://www.iea.org/publications/freepublications/publication/kwes.pdf.

Kasson, John. *Civilizing the Machine: Technology and Republican Values in America, 1776–1900.* New York: Hill and Wang, 1999.

Keizer, Garret. *The Unwanted Noise of Everything We Want.* New York: PublicAffairs, 2010.

Kelly, Brian. "Having It Their Way: Alabama Coal Operators and Their Search for Docile Labor, 1908–1921." In *It Is Union and Liberty: Alabama Coal Miners and the UMW.* Edited by Edwin Brown, 38–61. Tuscaloosa: University of Alabama Press, 1999.

Kern, Stephen. *The Culture of Time and Space, 1880–1920.* Cambridge, MA: Harvard University Press, 1983.

Kimbrell, Andrew, ed. *Fatal Harvest: The Tragedy of Industrial Agriculture.* San Rafael, CA: Foundation for Deep Ecology, 2002.

Kiple, Kenneth F., and Virginia Himmelsteib King. *Another Dimension to the Black Diaspora: Diet, Disease, Racism.* Cambridge, UK: Cambridge University Press, 1981.

Kirby, Jack Temple. *Mockingbird Song: Ecological Landscapes of the South.* Chapel Hill: University of North Carolina Press, 2008.

Kniffen, Fred, and Henry Glassie. "Building in Wood in the Eastern United States: A Time-Place Perspective." In *Material Culture Studies in America: An Anthology.* Edited by Thomas Schlereth, 237–250. Lanham, MD: Altamira Press, 2006.

LaDuke, Winona. "TVA's Big Plans for Navajo Uranium." *Business and Society Review* (March 1, 1979): 22–25.

Lears, T. J. Jackson. *No Place of Grace: Antimodernism and the Transformation of American Culture.* Chicago: Chicago University Press, 1994.

Lemenager, Stephanie. "The Aesthetics of Petroleum, after *Oil!*" *American Literary History* 24, no. 1 (Spring 2012): 59–86.

———. "Fossil Fuel: Manifesto for the Post-Oil Museum." *Journal of American Studies* 46 (2012): 375–394.

———. "Petro-Melancholia: The BP Blowout and the Arts of Grief." *Qui Parle* (Spring/Summer 2011): 25–56.

Livi-Bacci, Massimo. *A Concise History of World Population.* Malden, MA: Blackwell, 2007.

Lubowski, Ruben N., Marlow Vesterby, Shawn Bucholtz, Alba Baez, and Michael J. Roberts. "Major Uses of Land in the United States, 2002." US Department of Agriculture. *Economic Information Bulletin* No. EIB-14 (May 2006), http://www.ers.usda.gov/media/249896/eib14_reportsummary_1_.pdf.

MacCleery, Douglas. *American Forests: A History of Resilience and Recovery.* Durham, NC: Forest History Society, 2011, http://www.fia.fs.fed.us/library/briefings-summaries-overviews/docs/ForestFactsMetric.pdf.

Marchand, Roland. *Advertising the American Dream: Making Way for Modernity.* Berkeley: University of California Press, 1986.

Martschukat, Jürgen. "'The Art of Killing by Electricity': The Sublime and the Electric Chair." *Journal of American History* 89, no. 3 (December 2003): 900–921.

McAlister, Melanie. *Epic Encounters: Culture, Media, and U.S. Interests in the Middle East since 1945.* Berkeley: University of California Press, 2005.

McCusker, John J., and Russell R. Menard. *The Economy of British America, 1607–1789.* Chapel Hill: University of North Carolina Press, 1991.

McNeill, J. R. *Something New under the Sun: An Environmental History of the Twentieth-Century World.* New York: Norton, 2000.

Meikle, Jeffrey L. *American Plastic: A Cultural History.* New Brunswick, NJ: Rutgers University Press, 1995.

———. "Material Doubts: The Consequences of Plastic." *Environmental History* 2, no. 3 (July 1997): 278–300.

Menand, Louis. *The Metaphysical Club: A Story of Ideas in America.* New York: Farrar, Straus, and Giroux, 2002.

Mintz, Steven. *African American Voices: A Documentary Reader.* Oxford: Wiley-Blackwell, 2009.

Mosley, Stephen. *The Chimney of the World: A History of Smoke Pollution in Victorian and Edwardian Manchester.* Cambridge, UK: White Horse Press, 2001.

Mouhot, Jean Francois. "Past Connections and Present Similarities in Slave Ownership and Fossil Fuels." *Climate Change* 105, nos. 1–2 (March 2011): 329–355.

Mumford, Lewis. *The City in History: Its Origins, Its Transformations, and Its Prospects.* New York: Harcourt Brace Jovanovich, 1961.

———. *Technics and Civilization.* New York: Harcourt and Brace, 1963.

Nikiforuk, Andrew. *The Energy of Slaves: Oil and the New Servitude.* Vancouver, BC: Greystone Books, 2012.

———. *Tar Sands: Dirty Oil and the Future of a Continent.* Vancouver, BC: David Suzuki Foundation, 2009.

Nye, David. *America as Second Creation: Technology and Narratives of New Beginnings.* Cambridge: Massachusetts Institute of Technology Press, 2003.

———. *American Technological Sublime.* Cambridge: Massachusetts Institute of Technology Press, 1994.

———. *America's Assembly Line.* Cambridge: Massachusetts Institute of Technology Press, 2013.

———. *Consuming Power: A Social History of American Energies.* Cambridge: Massachusetts Institute of Technology Press, 1999.

———. *Electrifying America: Social Meanings of a New Technology, 1880–1940.* Cambridge: Massachusetts Institute of Technology Press, 1992.

Odum, Howard T. *Environment, Power, and Society.* New York: Wiley, 1971.

Olien, Roger, and Diana Olien. *Oil and Ideology: The Cultural Creation of the American Petroleum Industry.* Chapel Hill: University of North Carolina Press, 2000.

Orbanz, Eva, ed. *Filming Robert Flaherty's* Louisiana Story: *The Helen van Dongen Diary.* New York: Museum of Modern Art, 1998.

Ore, Janet. "Mobile Home Syndrome, Engineered Woods, and the Making of a Domestic Ecology in the Post–World War II Era." *Technology and Culture* 52, no. 2 (April 2011): 260–286.

Oswalt, Sonya, Mike Thompson, and W. Brad Smith, eds. *US Forest Resource Facts and Historical Trends.* US Department of Agriculture, Forest Inventory and Analysis National Program, 2009, http://fia.fs.fed.us. Accessed August 21, 2013.

Phillips, Sarah. *This Land, This Nation: Conservation, Rural America, and the New Deal.* New York: Cambridge University Press, 2007.

Pimentel, David, and Marcia Pimentel. *Food, Energy, Society.* New York: CRC Press, 2008.

Pollan, Michael. *Omnivore's Dilemma: A Natural History of Four Meals.* New York: Penguin, 2006.

———. "Power Steer." *New York Times Magazine*, March 31, 2002.

Pomeranz, Kenneth. *The Great Divergence: China, Europe, and the Making of the Modern World Economy.* Princeton, NJ: Princeton University Press, 2000.

Rabinbach, Anson. *The Human Motor: Energy, Fatigue, and the Origins of Modernity.* Berkeley: University of California Press, 1992.

Radkau, Joachim. *Nature and Power: A Global History of the Environment.* Translated by Thomas Dunlap. Cambridge, UK: Cambridge University Press, 2008.

———. *Wood: A History.* Translated by Patrick Camiller. Cambridge, UK: Polity, 2012.

Rao, Vickram. *Shale Gas: The Promise and Peril.* Research Triangle Park, NC: Research Triangle Institute Press, 2012.

Reece, Eric. *Lost Mountain: A Year in the Vanishing Wilderness.* New York: Riverhead Books, 2006.

Robertson, Thomas. *The Malthusian Moment: Global Population Growth and the Birth of American Environmentalism.* New Brunswick, NJ: Rutgers University Press, 2012.

Russell, Edmund, James Allison, Thomas Finger, John K. Brown, Brian Balogh, and W. Bernard Carlson. "The Nature of Power: Synthesizing the History of Technology and Environmental History." *Technology and Culture* 52 (April 2011): 246–259.

Rydell, Robert. *All the World's a Fair: Visions of Empire at American International Expositions, 1876–1916.* Chicago: University of Chicago Press, 1984.

Sabin, Paul. *Crude Politics: The California Oil Market, 1900–1940.* Berkeley: University of California Press, 2005.

Santiago, Myrna. *The Ecology of Oil: Environment, Labor, and the Mexican Revolution, 1900–1938.* Cambridge, UK: Cambridge University Press, 2006.

Schatzki, Theodore R., and Wolfgang Natter. "Sociocultural Bodies, Bodies Sociopolitical." In *The Social and Political Body.* Edited by Theodore R. Schatzki and Wolfgang Natter, 5–10. New York: Guilford Press, 1996.

Schlosser, Eric. *Fast Food Nation: The Dark Side of the American Meal.* New York: Houghton Mifflin, 2001.

Scott, James C. *Seeing Like a State: How Certain Schemes to Improve the Human Condition Have Failed.* New Haven, CT: Yale University Press, 1998.

Sexton, Steven, David Zilberman, Deepak Rajagopal, and Gal Hochman. "The Role of Biotechnology in a Sustainable Biofuel Future." *AgBioForum* 12, no. 1 (2009): 130–140, http://www.agbioforum.org.

Sieferle, Rolf Peter. *The Subterranean Forest: Energy Systems and the Industrial Revolution*. Translated by Michael P. Osman. Cambridge, UK: White Horse Press, 2001.

Singal, Daniel Joseph. "Modernism." In *A Companion to American Thought*. Edited by Richard Wightman Fox and James T. Kloppenberg, 460–462. Oxford, UK: Blackwell, 1998.

———. *The War Within: From Victorian to Modernist Thought in the South, 1919–1945*. Chapel Hill: University of North Carolina Press, 1982.

Smil, Vaclav. *Energies*. Cambridge: Massachusetts Institute of Technology Press, 1999.

———. *Energy in World History*. Boulder, CO: Westview Press, 1994.

———. "Nitrogen and Human Food Production: Proteins for Human Diets." *Ambio* 31, no. 2 (March 2002): 126–131.

Stallybrass, Peter, and Allon White. *The Politics and Poetics of Transgression*. Ithaca, NY: Cornell University Press, 1986.

Steinberg, Theodore L. "An Ecological Perspective on the Origins of Industrialization." *Environmental Review* 10, no. 4 (Winter 1986): 261–276.

Sutter, Paul. "The World with Us: The State of Environmental History." *Journal of American History* 100, no. 3 (2013): 94–119.

Szeman, Imre. "Crude Aesthetics: The Politics of Oil Documentaries." *Journal of American Studies* 46 (2012): 423–439.

Tichi, Cecelia. *Shifting Gears: Technology, Literature, Culture in Modernist America*. Chapel Hill: University of North Carolina Press, 1987.

Tobey, Ronald C. *Technology as Freedom: The New Deal and the Electrical Modernization of the American Home*. Berkeley: University of California Press, 1996.

US Department of Agriculture. *Livestock Slaughter 2011 Summary*. National Agricultural Statistics Service, April 2012, http://usda.mannlib.cornell.edu/MannUsda/viewDocumentInfo.do?documentID=1097.

———. *Poultry Production and Value 2011 Summary*. National Agricultural Statistics Service, April 2012, http://usda.mannlib.cornell.edu/MannUsda/viewDocumentInfo.do?documentID=1130.

US Department of Commerce. Bureau of the Census. *Historical Statistics of the United States: Colonial Times to 1970*. White Plains, NY: Kraus International Publications, 1989.

US Department of Energy. *Energy in the United States: A Brief History and Current Trends*, http://www.eia.gov/totalenergy/data/annual/previous.cfm#1999. Accessed February 20, 2013.

———. *Renewable Energy as Share of Total Primary Energy Consumption 2010*, October 19, 2011, http://www.eia.gov/totalenergy/data/annual/perspectives.cfm. Accessed January 30, 2012.

US Energy Information Administration. "How Much Oil Is Used to Make Plastic?" http://www.eia.gov/tools/faqs/faq.cfm?id=34&t=6. Accessed on February 27, 2013.

————. "How Much Petroleum Does the US Import and from Where?" http://www.eia.gov/tools/faqs/faq.cfm?id=727&t=6.

————. "Primary Energy Overview, 1949–2011." *Annual Energy Review.* September 27, 2012, http://www.eia.gov/totalenergy/data/annual/showtext .cfm?t=ptb0101.

————. "Primary Energy Use by Source and Sector, 2012." *Monthly Energy Review,* April 2013, http://www.eia.gov/energy_in_brief/article/major_energy _sources_and_users.cfm

————. "Table 1.1: Estimated Primary Energy Consumption in the United States, Selected Years, 1635–1945," September 27, 2012, http://www.eia .gov/totalenergy/data/annual/showtext.cfm?t=ptb1601. Accessed February 20, 2013.

————. "Table 1.3: Primary Energy Use by Source, 2012." *Monthly Energy Review,* March 2012, http://www.eia.gov/energy_in_brief/article/major_energy _sources_and_users.cfm.

————. "Table 1.5: Energy Consumption, Expenditures, and Emissions Indicators Estimates, Selected Years, 1949–2011." *Annual Energy Review 2011,* http://www.eia.gov/tools/faqs/faq.cfm?id=85&t=1. Accessed February 22, 2013.

————. "U.S. Primary Energy Consumption Estimates by Source, 1775–2011." *Annual Energy Review,* http://www.eia.gov/totalenergy/data/annual/per spectives.cfm. Accessed August 19, 2013.

US Environmental Protection Agency. "The EPA and Food Security," http://www.epa.gov/pesticides/factsheets/securty.htm. Accessed February 27, 2013.

US Geological Survey. "Iron and Steel Statistics, 1900–2009," November 29, 2012, http://minerals.usgs.gov/ds/2005/140/ds140-feste.pdf. Accessed August 19, 2013.

Vietor, Richard H. K. *Energy Policy in America since 1945.* Cambridge, UK: Cambridge University Press, 1984.

Vitalis, Robert. *America's Kingdom: Mythmaking on the Saudi Oil Frontier.* Palo Alto, CA: Stanford University Press, 2007.

Walker, Charles Rumford. *The Man on the Assembly Line.* Cambridge, MA: Harvard University Press, 1952.

Watts, Michael. "The Anatomy of a Petro-Insurgency in the Niger Delta." In *Crude Domination: An Anthropology of Oil.* Edited by Andrea Behrends, Stephen P. Reyna, and Günther Schlee, 49–70. New York: Berghahn Books, 2011.

————. "A Tale of Two Gulfs: Life, Death, and Dispossession along Two Oil Frontiers." *American Quarterly* 64, no. 3 (September 2012): 437–467.

Weiss, Thomas. "U.S. Labor Force Estimates and Economic Growth, 1800–1860." In *American Economic Growth and Standards of Living before the Civil War.* Edited by Robert E. Gallman and John Joseph Wallis, 19–78. Chicago: University of Chicago Press, 1992.

West, Elliot. *Contested Plains: Indians, Goldseekers, and the Rush to Colorado.* Lawrence: University Press of Kansas, 1998.

White, Leslie A. *The Evolution of Culture: The Development of Civilization to the Fall of Rome.* New York: McGraw-Hill, 1959.

———. "Energy and the Evolution of Culture." *American Anthropologist* 45, no. 3 (September 1943): 335–356.

White, Richard. *The Organic Machine: The Remaking of the Columbia River.* New York: Hill and Wang, 1992.

Worden, Daniel. "Fossil-Fuel Futurity: Oil in *Giant*." *Journal of American Studies* 46 (2012): 441–460.

World Bank, "Arable Land (Hectares per Person)," http://data.worldbank.org/indicator/AG.LND.ARBL.HA.PC. Accessed February 22, 2013.

Wrigley, E. A. *Continuity, Chance, and Change: The Character of the Industrial Revolution in England.* New York: Cambridge University Press, 1988.

Yanarelli, Ernest J., and Herbert G. Reid. "From 'Trained' Gorillas to 'Humanware': Repoliticizing the Body-Machine Complex between Fordism and Post-Fordism." In *The Social and Political Body.* Edited by Theodore R. Schatzki and Wolfgang Natter, 181–229. New York: Guilford Press, 1996.

Yergin, Daniel. *The Prize: The Epic Quest for Oil, Money, and Power.* New York: Free Press, 1992.

Young, Paulette. *Annual Coal Report 2011.* US Energy Information Administration, November 2011, http://www.eia.gov/coal/annual/pdf/acr.pdf. Accessed August 21, 2013.

Index